T0406322

Investor Relations and ESG Reporting in a Regulatory Perspective

Poul Lykkesfeldt · Laurits Louis Kjaergaard

Investor Relations and ESG Reporting in a Regulatory Perspective

A Practical Guide for Financial Market Participants

Poul Lykkesfeldt
Copenhagen, Denmark

Laurits Louis Kjaergaard
Copenhagen, Denmark

ISBN 978-3-031-05799-1 ISBN 978-3-031-05800-4 (eBook)
https://doi.org/10.1007/978-3-031-05800-4

Cover illustration: america365

This Palgrave Macmillan imprint is published by the registered company Springer Nature Switzerland AG
The registered company address is: Gewerbestrasse 11, 6330 Cham, Switzerland

About the Authors

Poul Lykkesfeldt is a managing partner, senior adviser and founder of Reliance A/S (https://reliance.dk), a leading Danish financial PR and communications firm with an international client base. Poul has over 35 years of experience in the fields of investor relations, financial transactional PR, investor activism, corporate/ESG communication and media relations related to listed and private companies, and private equity firms. Poul has previously served as head of IR, and deputy head of corporate communications, of Novo Nordisk; as a top-rated senior equity analyst with ABN AMRO; and as an investment banker and corporate finance adviser in London (UBS) and Copenhagen (KPMG Corporate Finance). Today, Poul advises international and domestic clients within financial PR/IR in connection with transactions, annual reports and ESG reports, as well as crises communications and media relations/training. Finally, Poul is a trusted adviser for company boards and the C-suite. Poul holds a M.Sc. (Econ.) from the Copenhagen Business School, Denmark and a PED from IMD, Switzerland.

Laurits Louis Kjaergaard is a strategy and business development manager at I&T, one of Denmark's most prominent independent asset managers. Having lived in six countries, Laurits has held finance and consulting positions at leading buy-side, sell-side, and regulatory advisory firms, including ABG Sundal Collier, PwC, Interogo (Inter IKEA Holding) and Danske Bank. During his career, Laurits has both worked with and prepared traditional company equity research, as well as ESG research reports, on listed companies in his earlier capacity as an equity research analyst. Prior to this, he advised companies on the implementation of selected EU legislation. Laurits holds a M.Sc. (Finance) from the University of Liechtenstein, a B.Sc. Business and Sociology from the Copenhagen Business School and a finance certification from the London School of Economics.

Testimonials

"All board members have a fiduciary duty to act on behalf of a company's shareholders and other stakeholders – this includes embracing today's significant developments within ESG reporting. This book provides an excellent and competent overview of the challenges that senior management and boards are facing today relating to new ESG reporting requirements. Further, it integrates the fields of ESG reporting, non-financial reporting, financial reporting and investor relations in a most helpful manner. It is evident that the authors are seasoned practitioners in their fields. This book is recommended as mandatory reading for board members; senior management; IR, ESG and communications executives; external advisers within ESG and IR; as well as relevant politicians; in respect to the ongoing journey of ESG reporting and the mobilisation of the UN's Sustainable Development Goals. I offer this book my highest recommendations as relevant and newsworthy reading regarding companies' and institutions' continued journey in pursuing the ESG agenda and meeting the increasing demands of investors regarding high-quality ESG reporting."

—Lise Kingo, *Independent Board Director in Sanofi SA, Covestro AG and Aker Horizons ASA. Former CEO and Executive Director of the United Nations Global Compact; and prior to this, Executive Vice President and member of Novo Nordisk's Executive Management*

"I doubt that a more insightful and practically-oriented handbook has been written, or may be written over the next decade, on the subject of investor relations, not only providing an essential overview but also illustrating how IR is executed in a best practice manner. This book integrates IR and ESG reporting, which is essential for global investors today. Further, this book cast light on companies' interaction and collaboration with investment banks and financial advisers, which is critical for any company - particularly small and medium-sized companies and in situations of an IPO, private equity or a venture capital journey. I strongly recommend this book to any board member, senior management, IR and ESG practitioner, as well as founder, IPO or capital raising candidate, who wish to increase their knowledge of the financial markets and appreciate how a company may effectively mobilise resources to satisfy its shareholders and stakeholders. This book also provides a competent 'down-to-earth' overview on relevant EU financial market regulation, as well as a checklist to increase transparency and reduce a company's risk premium to improve its capital structure and reduce its funding costs."

—Carsten Borring, *Associated Vice President Global Listing, Nasdaq. Member of The Federation of European Securities Exchanges (FESE) Listing Expert Group. Former board member of the Danish Centre for ESG Research and the Danish Government's Panel for Entrepreneurs*

"The EU regulatory wave towards the financial and capital markets has been significantly increasing over the past 10-15 years. Understanding the legal framework is a prerequisite for boards, senior management and company executives to navigate as a listed company. Failing to comply may have longstanding negative consequences. This book goes beyond simple regulatory analysis and provides an accessible and balanced overview of the increasingly complicated regulatory framework on the EU financial markets in terms of listing obligations, as well as financial and non-financial reporting, which any business leader or financial market practitioner may benefit greatly from. Further, this book provides a useful overview of the future trends in the EU regulatory area of financial and non-financial reporting, also in a global regulatory context. I take pleasure recommending this book to any stakeholder with an interest in the EU regulatory framework related to the financial and capital markets, including to companies and investors outside the EU who wish a competent EU regulatory overview."

—David Moalem, *Partner, Ph.D., Bech-Bruun law firm, Capital Markets & Financial Institutions Team. Former Partner and National Head of Legal Services at Deloitte, and has held a wide range of official appointments in the Danish government's and Danish FSA's working groups for EU financial regulation*

"The authors provide unique and incredibly insightful practical information and recommendations for IROs and Executive Management teams in listed companies regarding organising and executing best practice IR. The combination of simple and straightforward information, and its many layers of solid theoretical understanding and real-life examples, makes this book invaluable for both the experienced and the newly started IRO. It serves as a practical checklist in the daily dealings with investors and equity analysts. This includes optimising the resources of IR and IR performance; creating an IR department; overview of relevant IR tools; the increasing importance of retail investors and digital IR; and illustrating the execution of perception studies among equity analysts and institutional investors which the authors refer to as perhaps the most underrated IR tool. Finally, this book provides an excellent detailed guide to creating a takeover response manual and considerations regarding investor activists, including the use of external advisers; the implications of relevant major EU regulations; and how to practically conduct ESG reporting, both for beginners in the field, and for experienced companies. This book is, in every manner, highly recommended."

—Michael Bjergby, *Senior Vice President, Head of Group Finance, ISS A/S. Former Head of Investor Relations at ISS, Pandora and GN Group*

Preface

In recent years, European Union (EU) regulators have introduced legislation to increase transparency and investor protection in the financial markets. In particular, this has impacted investment banks through the implementation of MiFID II and MiFIR in 2018. The focus has since advanced to institutional investors and companies to integrate stakeholder capitalism and ESG (environmental, social and governance) factors alongside the quest for profit.

The EU is expected to invest and mobilise 1 trillion euro in its Green New deal from 2021 to 2027 to kick-start green investing to pursue the UN's Sustainable Development Goals (SDGs) and the science-based emission targets.[1] However, a common language of sustainability is needed to mobilise capital efficiently. Therefore, for the next ten years, the EU's taxonomy and associated legislation (CSRD, SFDR and updated MiFID II) will be implemented to provide a tailwind for non-financial reporting. The quest is for companies to embrace strategic ESG in the companies' operations and investors' strategies to pursue a sustainable economy.

In our work, we have witnessed a clear need for financial market practitioners to fully understand the most profound regulatory framework to ensure compliance and embrace new opportunities in a competitive market. Many institutional investors and investment banks are currently building in-house ESG equity research capabilities—and companies are hiring ESG experts to support their journey in the field. However, we believe these are trivial moves if the board of directors and senior management do not properly embrace the regulatory framework of ESG with a view to share the vision of ESG. It is a fiduciary responsibility of board members to represent the company's investors—and this now includes best-in-class both financial and non-financial performance transparency. From all directions, listed and unlisted companies will be scrutinised. They will, in the future, either pass or fail the test based on their approach to non-financial reporting and mobilising ESG in their existing business strategy and processes.

All major companies in the EU (listed and unlisted) must implement non-financial reporting (named CSRD) from 1 January 2023—and all companies must likely do so from 1 January 2026. In addition, all major institutional investors marketing towards EU investors must disclose their ESG strategy and performance (applicable from 10 March 2021 named SFDR) on an entity level from 30 June 2021 and on a product level from 1 January 2022. All professional investors must likely do so from 1 January 2025. These must be updated annually according to its accounting reference periods. Major investors must also gather, consider and integrate their client's sustainability preferences from 2 August 2022 (amendment to MiFID II). Their clients and the regulators will pressure institutional investors to embrace ESG—who in turn will further pressure companies to increase transparency and run their businesses with a purpose and consideration of its stakeholders.

This book does under no circumstances constitute a replacement of the advice of a companies' internal and external legal counsel in respect of the regulatory framework, but is instead a supplementary handbook and overview for board members and senior management of listed companies, unlisted companies and companies considering an IPO (initial public offering); investor relations, ESG and communications practitioners; institutional and private investors; private equity executives; investment bankers; legal practitioners; accountants and auditors; financial journalists; politicians; as well as university and business school students; to understand the dynamics of the financial markets and the direction they are moving in respect of ESG. Further, as both ESG as such and the regulatory framework related to non-financial/ESG reporting are far from matured and will continue to develop during the years to come, companies are encouraged to closely follow the developments in this area and seek to stay in the forefront, as the financial markets are believed to reward this behaviour in the longer term. Similarly, investors and equity analysts have a significant task in front of them adapting to the new ESG future and develop new quantifiable ESG-related valuation tools to be used in company valuation, as they for many decades solely have relied on valuation multiples and models where the input consists of financial information only.

Best-practice investor relations (IR) is essential for any listed company's ability to seek a low-risk premium from investors in order to utilise the advantages of the financial markets and secure an optimal capital structure. In addition, companies must—in parallel with its IR activities—embrace their stakeholders to ensure the company's long-term position in a globally competitive environment. We hope this book will provide a solid framework and a helping hand for companies to conduct best-practice IR and include best-in-class ESG approaches in their businesses.

It was fascinating to witness the debates in January 2022 at the Economic Forum Agenda in Davos, and the UN Climate Change Conference (COP26) in Glasgow in November 2021. They both attempted to make sense, debate and understand how to embrace the rising field of ESG and increase transparency. Notably, panel members at the Davos Agenda acknowledged how complicated the ESG area had become. Some even, as a humorous side comment, advised young people to commence studying non-financial/ESG reporting at university due to the significant shortage expected in the immediate future of ESG-orientated business leaders, advisors, lawyers and accountants/auditors. We hope this book will not only contribute to companies' and financial communities' understanding of non-financial/ESG reporting and illustrate that there is still a long journey ahead—but also serve as an inspiration, guidance and motivation for young people who are about to choose their career path.

We have known each other professionally for a handful of years. Quickly, our common interest for ESG materialised between us. Once this was established, we quickly decided to write this book to share our views and to seek to provide a helping hand to others with a wish to move forward the ESG agenda and the integration between financial and the extremely important non-financial reporting. We further decided, as we both have a deep-rooted passion and interest in the financial markets and their actors, to share our experiences regarding, e.g. IR, takeover response manuals and other communication contingencies, as well as companies' collaboration with investment banks and corporate finance advisers, in an attempt to provide other practitioners with insight into these issues, as these areas are interlinked and also highly relevant for boards of directors and senior management. Hence, these themes are all integrated in this book.

As we began on our book project a little over a year ago, we have during our work process witnessed the daily acceleration and increased momentum of news commentary on ESG and stakeholder capitalism, and how to navigate in the transforming financial markets. It is very encouraging to see the above-mentioned momentum rolling into the current "decade for action" (as referred to by the United Nations Global Compact) and we are excited that this book may potentially present our small contribution to the ever so important development in the global ESG agenda, both to companies and to the participants of the financial market.

We are truly humbled and proud to author this book and hope that our readers will appreciate it.

Copenhagen, Denmark

Poul Lykkesfeldt
Laurits Louis Kjaergaard

Note

1. Eurocoop. (2020). Green deal and farm to fork strategy, https://www.eurocoop.coop/news/282-DIGEST-Green-Deal-and-Farm-to-Fork-Strategy.html.

Acknowledgements

We would like to extend our deep-rooted gratitude to the four esteemed experts and reviewers who, in spite of their already very busy schedules, took the time to share their insights in discussions with us and to offer valuable input to our draft manuscript. We feel truly blessed that they without hesitation all committed themselves to become involved in this project.

As we from the beginning had high ambitions for the content of this book, we had a great wish that it could stand the test of their constructive comments, challenges and advice with regard to the book's four main themes:

- The financial markets with a focus on the stock market.
- Investor relations (IR).
- Non-financial/ESG reporting.
- The regulatory framework in the EU related to the financial markets, including non-financial/ESG reporting, as well as global trends in the area.

On this background, we would like to provide the following supplementary gratitude and comments to our four reviewers (in the order to the above-mentioned main themes of the book):

Carsten Boring, Associated Vice President Global Listing, Nasdaq
With 20 years at Nasdaq, and a further background in corporate finance and the capital markets, Carsten has been deeply involved in the immense transformation of the financial markets during that time. Further, Carsten has over recent years been a pioneer for the sustainable acceleration of listed companies and their quest to integrate ESG reporting in the companies' overall financial reporting, both via Nasdaq and his role as a former board member of the Danish Centre for ESG Research. Finally, Carsten has also been instrumental in Denmark, as the architect behind the Nasdaq First North market,

for the development of the IPO market, in particular for European growth companies. As an ambassador for the financial community as a member of the Danish Government's Panel for Entrepreneurs and a member of the Federation of European Securities Exchanges (FESE) Listing Expert Group, Carsten has provided us with valuable feedback on our views on the financial markets as well as the expected future developments and dynamics of the stock exchange as a marketplace, which we thank him for.

Michael Bjergby, Senior Vice President, Head of Group Finance, ISS A/S

Michael has an impressive resumé in the investor relations field as former head of IR of three of Denmark's most prominent companies and global leaders in their field: ISS, Pandora and GN Group. Further, as former Group CFO of HTL-Strefa, he has both conducted IR from a senior management perspective as well as been part of a private equity exit. His diligent and positive attitude has provided us with extra motivation and inspiration on our journey of authoring this book. We would like to thank Michael for his feedback in the IR field and as to how ISS, as the world's largest service company and one of the world's biggest employers, embraces ESG and practically transforms its approach towards non-financial measures and ESG reporting.

Lise Kingo, Independent Board Director in Sanofi SA, Covestro AG and Aker Horizons ASA

Lise is a globally acknowledged pioneer in the ESG area. At her time with Novo Nordisk, the leading global healthcare and diabetes company, she was responsible for publishing one of the first environmental reports in the world in 1994. She was the main driver behind Novo Nordisk's "triple bottom line" reporting which a few years later led to the company's full integration of financial and non-financial reporting. These steps are widely seen as laying the ground for today's ESG reporting globally. In 2002, Lise was promoted to Executive Vice President and member of Novo Nordisk's Executive Management. Through the years 2015–2020, Lise was CEO and Executive Director of the United Nations Global Compact which introduced the SDGs to its more than 13,000 members globally and contributed to making ESG a strategic, boardroom priority. We are extremely thankful to Lise for our valuable discussions and her significant contribution to our own observations regarding the current and future developments in the ESG area, including non-financial reporting. Further, we very much appreciate Lise for sharing her time and thoughts on how to implement ESG from the board level and the importance of stakeholder capitalism. Lise's most powerful message is that ESG must be anchored first at the strategic level in the company's purpose, governance and corporate business strategy, then cascaded across the organisation and finally communicated to all stakeholders.

David Moalem, Partner, Ph.D., Bech-Bruun law firm, Capital Markets & Financial Institutions Team
We are extremely fortunate to have attracted the attention and interest of David for our book project. As a partner at the Danish market-leading law-firm Bech-Bruun, and the most credited Danish lawyer and legal expert in Danish as well as EU financial and capital markets regulation, cf. e.g. Legal500, Chambers, IFLR1000, etc., combined with his many official appointments in the Danish government's working groups for EU financial regulation, and being the author of around 60 books and articles on EU and Danish legislation in respect of the financial and capital markets, we have had the support of an incredible capacity in the EU legal field. We are most grateful for David's time and for sharing his comments, insights and advice regarding our interpretation and mapping of the current EU regulatory framework regarding the capital markets. This also goes for the coming trends in EU laws and regulations in both the financial and non-financial/ESG reporting area—all with a view to provide the reader with an overview of the current status and future developments.

In addition to our four reviewers of the book, we would like to extend our thanks to Palgrave Macmillan, especially, Tula Weis and Supraja Ganesh, for the support, as well as for a pleasant and professional process throughout the preparation of this book.

Finally, to our loved ones, family and friends, we ask for forgiveness for during this project having dived into the abyss for days on end and into the late hours. You have given us more motivation and energy than you may possibly imagine.

Introduction

The financial markets, in which the financial industry interacts, are the world's largest marketplace and serve as a critical pillar in our society. Despite being tackled by monumental changes in regulatory, technological and socio-economic forces, there is likely no other industry that has been as transformative when faced with game-changing challenges. Yet, despite its transformative behaviour, society seems no longer to view the financial markets as the aggregation of willing buyers and sellers exchanging securities at a fair market price. Instead, especially in the aftermath of the 2008 financial crisis, the financial markets are viewed as a dishonourable capitalist anti-Robin Hood type system where sophisticated major institutions come out ahead on behalf of everyone else.

Regulators have recognised this and aim to increase the protection of financial markets participants, protect the integrity of the markets and promote fair, transparent, efficient and integrated markets for all stakeholders. Following the financial crisis, the European Union (EU) initiated a series of directives and regulations to support their aim and coined the "regulatory wave". This wave has continued and today moved from the integrity of the markets to embracing stakeholder capitalism and including environmental, social and governance (ESG) standards into policy. The EU has clearly become the front-runner in setting ESG standards, and we believe many countries will follow its lead. The core motive is to include the purposes of people, prosperity and the planet alongside a company's quest for profits. Having an integrated ESG approach to business is not only important, but ultimately a license to operate.

At the 2022 World Economic Forum in Davos, the main topic was ESG metrics for a sustainable future. This is in light that all major EU institutional investors and companies are obligated to disclose their approach and performance of ESG from 1 January 2022 and 1 January 2023, respectively. However, it was clear at the summit that the confusion of non-financial/ESG reporting is considerable, and more information and consolidation are needed.

Nevertheless, the distinguished panel members' conclusion recognised the tremendous speed at which ESG is developing and the associated best practice reporting standards.

As it is impossible for us to focus on all topics on the dynamic and important field of non-financial/ESG reporting, this practical handbook focuses on the financial markets from the perspective of equities and listed companies, i.e. the stock markets. However, it is evident that there is also a huge and increasing need to include non-financial/ESG considerations in connection with the UN's mission of its Sustainable Development Goals (SDGs) in corporate financing, i.e. bonds and the capital markets. On this matter, we wish to pass on the baton on this important topic and hope that our book may inspire other practitioners to explore this important topic in more detail.

The financial market is socially constructed, which means that it requires social coordination, consensual validation and stakeholder management to limit the risk premium of a company's stock price and its ability to raise capital. Therefore, managing investor relations (IR) plays a vital part to support a listed company's strategy; communication; as well as the management of expectations and capital structure; considering the pressures from the financial markets. Therefore, we do not confine the IR role to a specific investor relations officer (IRO), but any IR relation, process and interaction that the company may have with the financial community and the non-financial stakeholders. Nevertheless, for reasons of simplicity and terminology, we refer in this book to "the IRO" when referring to the holistic interaction between the company and the financial markets.

Given our practical insights in the financial markets from the perspectives of investors, equity analysts, regulators and IR, we seek to uncover the monumental changes of the financial markets and the associated working conditions of its actors. We aim to provide a practitioner's framework to encapsulate the differences appropriately in a practical format.

With our combined experiences as current practitioners, we seek to uncover relevant IR and financial communication areas and how to deal with them. The requirements for good IR and financial communication, as well as good corporate governance, have increased significantly over the past 20 years. The companies that do not prioritise this are increasingly punished in the form of low interest from investors and equity analysts. This usually leads to less liquidity in the company's shares and a lower share price, which creates a less optimal capital structure and higher funding costs and increases the risk of attempts to take over a company. In the short and medium term, foundational or majority-dominated ownership may reduce this risk, but hardly in the longer term due to possible changes in legislation or pressure from investors.

The financial community, regulators and other stakeholders demand good non-financial/ESG reporting. Financial data and information are essential for any IRO; however, they do not render the complete picture. We need to look further ahead.

The role of corporations in society has never been more critical than during the global turbulence caused by the COVID-19 pandemic. Old virtues such as trust, decency and responsibility demonstrated these days are decisive for companies' future sales and attracting investors. The financial markets, for good reasons, always consider future risks and opportunities. Institutional investors search for long-term growth potential and responsible value creation. The professional participants of the financial markets call for transparency. So, if companies wish investors to gain insight into relevant ESG policies, targets and results, they need to professionalise how they work with and communicate non-financial disclosure.

Consistent and standardised figures, methods and definitions form the basis of transparent and robust ESG reporting. When presented with an ESG report with a convincing communicative edge to it, the ESG profile will be a strong point for a business. On this background, all companies are advised to commence the progress of evaluating its ESG standpoints and look at its business in a relevant context, with a view of taking its ESG reporting to the next level or get properly started.

In this book, we further outline the major regulatory themes related to the European financial markets and put them into perspective with IR and ESG reporting—and at the end of the book, in a both a global and future context.

This book contains ten sections. Firstly, we address how the financial markets work (parts 1–2), thereafter, the EU's financial regulatory environment (part 3), then the practicalities and dynamics of IR (parts 4–5), and finally, we discuss implementing and improving non-financial/ESG reporting, as well as embracing stakeholder capitalism, and what the future may bring in terms of non-financial/ESG reporting (parts 6–10):

- Part 1 aims to uncover how the financial markets work, the motivation for companies to be listed and which variables are included in the pricing of the companies' shares on the stock exchange. We have also shed light on how ESG factors are implemented in traditional equity research and analysis.
- In part 2, we discuss the ecosystem of the financial markets, including their respective actors, their motivation, responsibilities and how they are intertwined relative to the role of the IRO. We believe it is essential for any IRO to have a solid understanding of how investors (institutional and retail) and investment banks (and brokers) are organised to communicate effectively with them.
- Part 3 outlines the major regulatory themes related to the European financial markets. This includes how the EU regulates the financial markets and an overview of the regulatory wave of MiFID II/MiFIR/MAR following the 2008 financial crisis. We also outline the new non-financial/ESG reporting legislation of taxonomy, SFDR and CSRD, which is currently being implemented. Finally, we discuss the

main themes relevant for the IRO and mitigation tools of the already implemented legislation.

- Part 4 focuses on best-practice IR, given how the financial markets work, its participants and the regulatory environment. Next, we discuss the purpose and functional role of IR. Finally, we focus on practical IR, including IR tools, requirements and thoughts on managing expectations and on embracing digital IR.
- Part 5 discusses IR's role in takeover response situations, contingency planning, investor activism, external advisers (including corporate finance, lawyers, accountants, IR consultants and proxy advisers), shareholder engagement, crisis communication, as well as considerations regarding investor activists and initial public offering (IPO).
- We outline in part 6 how a company's stakeholders are embracing ESG, COVID-19's impact, and the UN's SDGs as the underlying basis and political motivation of non-financial legislation.
- In part 7, we outline the unconsolidated market of non-financial/ESG reporting standards. Finally, considering the standards, we list how the current non-financial frameworks are intertwined relative to ESG criteria and discuss the level 1–3 scopes of taxonomy.
- Part 8 sets the agenda and provides recommendations for the company's first non-financial/ESG report in four stages—establish, expand, embed and enhance. We focus on mobilising resources and include thoughts about incentive structures to onboard the organisation.
- In part 9, we continue from section eight on enhancement. We provide follow-up ideas and inspiration for the company to consider delivering and thriving towards best-in-class ESG reporting.
- Finally, in part 10, we share our thoughts regarding the future of reporting trends—from moving beyond gross domestic product (GDP) and profits to embracing stakeholder capitalism of people, prosperity and the planet. We believe best-practice IR needs to mobilise the organisation to embrace and explore future business opportunities.

In a brief postscript, we seek to summarise our considerations and conclusions in a global perspective.

It is our hope that a wide range of business professionals, institutional and retail investors, politicians and students at universities and business schools will find inspiration in this book based on our insights into the day-to-day operations of the different stakeholders. However, our book is primarily meant as a guiding tool aimed at members of the board of directors and senior management, and those responsible for, and practising, IR, ESG, communication, law and finance with and within companies.

The final manuscript for this book was submitted to the publisher on 1 March 2022. Therefore, the book does not include any comments to the regulatory developments following this date, nor does the book include any comments regarding the consequences on the financial markets of Russia's invasion of Ukraine at the end of February 2022.

Contents

List of Figures

Abbreviations

AGM	Annual General Meeting
AML	Anti-Money Laundry
AUM	Assets Under Management
B2B, B2C and B2G	Business to Business, Business to Consumers and Business to Government
BRT	Business Roundtable
BS	Balance Sheet
BU	Business Unit
CA	Corporate Access
CAPEX	Capital Expenditures
CDSB	Climate Disclosure Standards Board
CEO	Chief Executive Officer
CF	Cash Flow Statement
CFO	Chief Financial Officer
CMD	Capital Market Day
CoI	Conflict of Interest
CSR	Corporate Social Responsibility
CSRD/NFRD	Corporate Sustainability Reporting Directive/Non-Financial Reporting Directives
DCF	Discount Free Cash Flow
EAMP	Emission Allowance Market Participant
EBITDA	Earnings Before Interest, Tax, Depreciations and Amortisation
ECM/DCM	Equity Capital Markets and Debt Capital Markets
ECP	Eligible Counterparties
EGM	Extraordinary General Meeting
ESAP	European Single Access Point
ESG	Environmental, Social and (Corporate) Governance
ESMA	European Securities and Market Authority
ETF	Exchange-Traded Fund
EU	European Union
EV	Enterprise Value
FCF	Free Cash Flow

FHAF	Federal Housing Financing Agency (the US)
FSB	Task Force on Climate-Related Disclosures
FSB	The Federal Reserve Board (the US)
FSOC	Financial Stability Oversight Council (the US)
GAO	Governance Accounting Office
GDP	Gross Domestic Product
GRI	Global Reporting Initiative
IBC	International Business Council
IIRC	International Integrated Reporting Council
IMC	Issue Management Contingency
IPO	Initial Public Offering
IR and IRO	Investor Relations and Investor Relations Officer
ISSB	International Sustainability Standards Board
ITF	Intention to Float
KPI	Key Performance Indicator
KYC	Know Your Customer
M&A	Mergers and Acquisitions
MAR	Market Abuse Regulation
MDGs/SDGs	Millennium Development Goals and Sustainable Development Goals
MiFID II	Market in Financial Instrument Directive
MiFIR	Market in Financial Instrument Regulation
NGO	Non-Government Institution
OPEX	Operating Expenses
OTC	Commodity Futures Trading Commission (the US)
OTF/MTF	Organised Trading Facility and Multilateral Trading Facility
P&L	Profit and Loss Statement
PAI	Principal Adverse Impact Factors
PDMR	Personal Discharge Managerial Responsibility
PESTEL-DC	Political, Economic, Social, Technological, Environmental (and ethics), Legal, Demographical, Competition
PM	Portfolio Manager
PR	Public Relations
PtX	Power-to-X
Q&A	Questions and Answers
R&D	Research and Development
ROCE	Return on Capital Employed
SASB	Standards of the Sustainability Accounting Standards Board
SBTi	Science-Based Targets Initiative
SDR	Sustainable Disclosure Requirements (UK)
SEC	Securities and Exchange Commission (the US)
SFDR	Sustainable Finance Disclosure Regulation
SMART	Specific, Measurable, Achievable, Realistic and Time-Related
SME	Small-Medium Enterprises
SPAC	Special Purpose Acquisition Company
SROI	Social Return on Investments
TCFD	Task Force on Climate-Related Disclosure
TRM	Takeover Response Manual
UN	United Nations
UNGC	United National Global Compact
WEF	World Economic Forum

PART I

The Financial Markets: An Overview

CHAPTER 1

The Benefits and Drawbacks of a Stock Market Listing

To be a successful business owner and investor you must be emotionally neutral to winning and losing. Winning and losing is just part of the game.

—Robert T. Kiyosaki, Author

What Are the Financial Markets?

The classical rule of communication is "know your audience". Most board members, members of the senior management, founders, shareholders of companies and investor relations (IR) representatives are general practitioners with distinctive backgrounds. Therefore, we initially intend briefly to highlight the underlying dynamics of the financial markets to uncover how to practice best-practice IR and how it can create value for anyone without a distinctive background in finance, and to uncover the main benefits and drawbacks of a listing.

Today, global financial assets are worth 370 trillion euro (422 trillion US dollar).[1] Therefore, the financial markets are the most significant marketplace in the world. It is a function where buyers and the sellers meet to trade financial securities, instruments and products. In the past, transactions were conducted physically; however, with the digital revolution, nearly all transactions are today digital. A wide range of financial securities, instruments and products are traded on the exchanges from equities (also known as "stocks"), bonds and other debt-related securities ("the capital markets"), commodities, currencies, real estate, indices and financial instruments based on these securities. A transaction occurs when a seller and buyer agree on a price on financial security, which happens 6–10 billion times every day in the US alone.[2]

P. Lykkesfeldt and L. L. Kjaergaard, *Investor Relations and ESG Reporting in a Regulatory Perspective*, https://doi.org/10.1007/978-3-031-05800-4_1

This book will focus on the market for equities, defined as publicly traded shares of companies on open stock exchanges. Equities are essentially a share of a company traded among investors such as institutional investors (pension funds, hedge funds, insurance companies, etc.), retail investors, companies and governments. To ensure a fair transaction process, parties that engage in a transaction must have a high level of trust and confidence in the system and believe that there is fairness in all transactions that they engage in. As a result, the financial markets are today highly regulated to counter misuse and cheating, and to ensure investor protection and transparency while safeguarding financial stability.

Positive Aspects of a Stock Listing

It requires an in-depth due diligence and application process; however, any private company can choose to seek a listing of its shares on a publicly traded stock exchange, where the main benefits are numerous (Fig. 1.1).

For any of these benefits to have a viable effect, a listed company needs to act in a transparent (yet balanced) manner towards its shareholders and stakeholders; this optimises the probability of high trustworthiness and a stable developing share price that reflects the fundamental direction of the business. In addition, a company should aim to improve the shares' stability and counter the threat of being held hostage to a particular group of investors. Therefore, a company needs to facilitate and target a broad and diverse investor base with differing motives, horizons, risk appetite and needs. This is important because it is a fundamental motivation for a strong IR function.

Easier Access to Capital

The neo-capitalist economic purpose of a company in a capitalist society is to combine investors' capital and labour as an input used in producing goods or services to generate an economic profit for the shareholders in the company. Therefore, the company needs to attract favourable financing of its capital needs ("a low cost of capital") and competent employees at attractive costs ("wages" or "salaries") to deliver a competing product or service.

Suppose the company can deliver a competing product or service. In that case, it can increase its sales ("sold volumes and/or price") and become more cost-effective through economies of scale thereby increasing profit margins, while sustainably expanding its business and thus delivering an attractive level of cash flow to its shareholders. Due to the vast number of companies listed on the stock exchange, investors have high freedom of choice regarding where to allocate their capital. Therefore, an investor seeks to find investments where the expected return on the investments is reflected in the riskiness of the investment itself. Therefore, the riskier an investment, the higher the investor's expected return.

If a company can manage expectations in the long run, it lowers its risk profile. It allows investors to include assumptions about the future in their

Fig. 1.1 Positive aspects of a stock market listing

investment decision more fairly. This leads to more investors interested in allocating their capital to the company, thus increasing competition and demand among investors, leading to a broad investor base and more favourable and flexible financing terms. Therefore, it is in the company's interest to manage expectations about the future to allow easier access to capital and thus more financial flexibility to deliver a long-term economic profit. Managing expectations vis-à-vis investors and equity analyst is an art rather than a science and has always been a challenge. While the financial markets historically have focused on revenues, growth, margins, costs and CAPEX (capital expenditure), onwards ESG-related factors may be just as important.

Liquidity in the Ownership

High liquidity of shares is often related to disturbance of financial assumptions, and therefore, higher riskiness of an investment lowers confidence from existing shareholders. A balanced volatility level (a low deviation in the average share price development) allows investors the flexibility to buy or sell shares more easily, lowering the overall investment risk. Therefore, it is in the interest of all investors and companies to have a balanced volatility level of traded volumes of shares supported by managing expectations and facilitating enough shares to be traded on the financial markets. Whereas the share value is priced on an ongoing basis on the stock exchange, the number of shares is relatively stable and dictated on the general assembly of a company's shareholders. A company can issue or buy back more shares, therefore controlling the number of shares available to investors.

Out of the total shares made available to investors in a company, often companies have one or several legacy investors (such as the founder(s)) and/or long-term financial institutional investors) in their investor base. The total number of outstanding shares less the shareholdings of the legacy shareholders is known as the "free float" and refers to the proportion of shares traded more regularly. A company can buy back these shares to lower the number of traded shares, thus providing a demand for the shares which may increase the share price, and thereby also reducing the volatility of the existing shares. On the other hand, the company can issue more shares, leading to an increase in the number of its existing shares and thus higher volatility. By issuing new shares, the company is essentially printing and selling shares and thus raising capital, but thereby also allocating future profits among more shareholders; this is advantageous if the company is looking to decrease debt or finance new operations, which are believed to provide a return on investment above the company's WACC (weighted average cost of capital).

Transparency and Servicing Minority Investors and Stakeholders

Private companies are typically owned by relatively few investors, often the founder(s) and their friends and family. They may also have onboarded other investors, including private equity funds, venture capital funds, or seed/angel investors that serve as early-stage capital for an entrepreneurial business idea. In such a set-up, the ownership structure is dictated by private arrangements and contracts among the existing owners. This naturally decreases transparency, especially among minority investors, who have less control of the company and insights into the agreements and contracts.

If the company is traded on a public stock exchange, high transparency requirements are implemented, allowing minority investors to have better insights into the company's operations and ownership structure. In addition, there is essentially only one market price in a stock listed setting, a fixed number of tradable shares and number of voting rights for the general assembly. Therefore, if a company has or aims to have a high proportion of minority investors, a listing on a stock exchange is a significant benefit to facilitate their interests.

Ease of Ownership Transition and Exit Route to Investors

A stock listed company allows for more favourable and flexible financing terms. This can be important if a founder or an early-stage investor wants to realise their gain, diversify their financial portfolio, transfer ownership to the heirs or simply exit the investment for other reasons.

These options are possible in a private set-up; however, with divesting to more minority investors (rather than relying on a few large investors or the existing investors), the owner will achieve more favourable and transparent terms from the divestment on the financial markets.

Credibility and Marketing

Trust and credibility are critical when engaging in new and ongoing business relationships with customers and suppliers. When engaging in these relationships, counterparties may conduct an ongoing due diligence process and balance business activity with the counterparty's inherent business and credit risk. In addition, the bargaining power of prices for a company towards its counterparties is significantly reduced if credibility is low as the risk is higher.

As a result of the high due diligence, transparency requirements and clear ownership communicated associated with becoming and being listed on a stock exchange, a successful listing itself typically serves as a powerful "blue-stamp". This strengthens the company towards its business counterparties, allowing for easier marketing and trading terms. The credibility of a listed company typically increases—versus its privately-owned competitors—as the company becomes more known and it shows consistent solid and transparent results.

Employee Engagement

We believe employee engagement is a less acknowledged benefit of being a stock listed company. Previously, we noted that it is fundamental for a company to have competent employees at attractive prices, i.e. wages. A company can offer employee ownership programmes to increase recruiting efforts of new employees and retain qualified talent.

This may essentially allow employees to purchase employee shares at attractive prices or remunerate employees with a share-based bonus programme. A company can thus increase engagement which increases the company's stability and therefore their business relationships and costs. However, rules regarding employee shares and share-based bonus programmes vary from country to country.

Negative Aspects of a Stock Listing

We believe that for a wide number of companies, a stock listing has clear benefits. A stock listing is optimal if a company seeks a diverse investor base, targets easier access to ongoing capital and has more accessible exit options for the existing shareholders. However, if this is not the case, it is less obvious to be stock listed. The board of directors, senior management and existing shareholders must, in such a case, consider the potential negative aspects of a stock listing which may include the ongoing direct and indirect costs of maintaining a listing, management and time resources related to IR activities and ensuring regulatory compliance, as well as a certain market pressure to focus on short financial (quarterly) performance (Fig. 1.2).

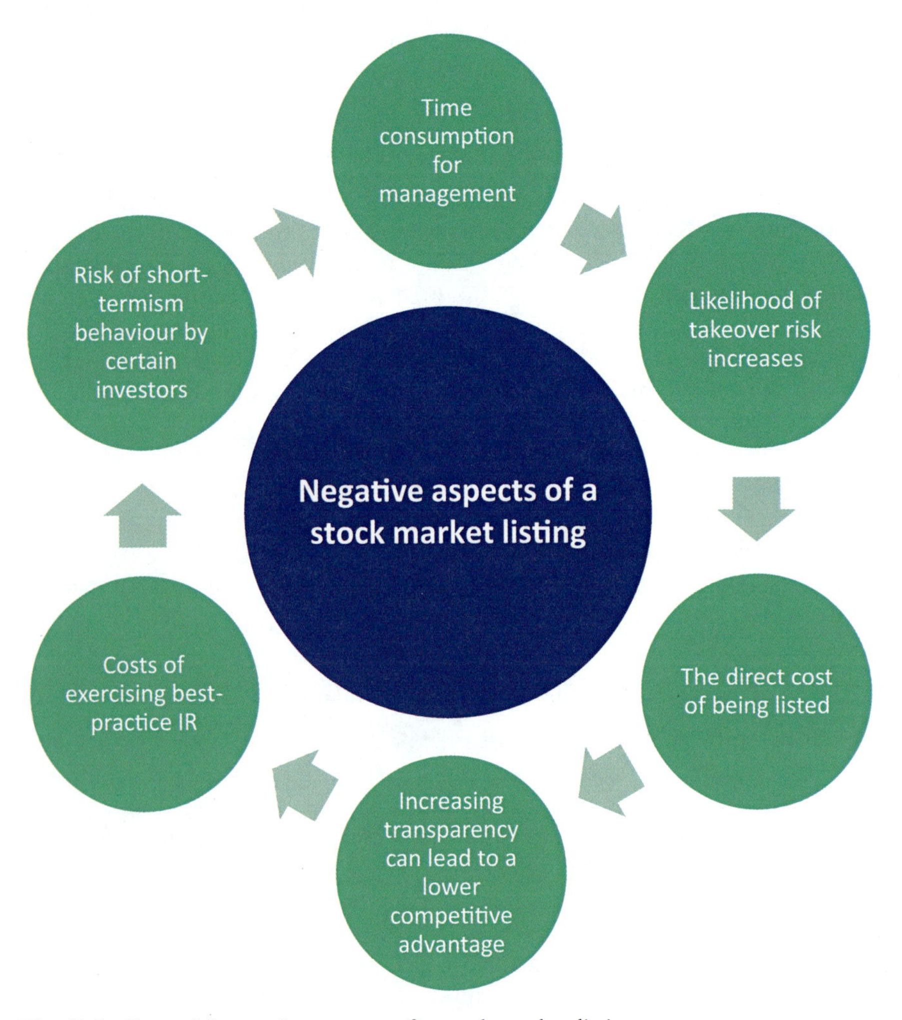

Fig. 1.2 Potential negative aspects of a stock market listing

References and Sources

1. Bahgai, P., et al. (2021). *Crossing the horizon: North American asset management in the 2020s*. McKinsey & Company.
2. Vlastelica, R. (2017). *US stock trading volume hit a three-year low in 2018 amid near-absent volatility*. MarketWatch.

CHAPTER 2

The Formation of Stock Prices

What Moves the Stock Market?

Suppose it was possible to determine a company's future cash inflow and outflow (together net cash flow) until it is extinct and discount back those cash flows at an appropriate risk-free discount at today's point in time. In that case, the intrinsic value of the stock would be known. However, in real life, it is unknown. Therefore, an investor must include the appropriate factors and uncertainties to estimate the cash inflow and outflow and the discounting factor. As a result, fundamental investors (also known as value investors) attempt to understand a business to determine roughly how they will perform in the next 10–30 years by setting assumptions of how competitive forces can shape the company, as well as what the demand of the company's products and services will be, including the company's ability to perform throughout a business cycle.

Fundamental investing is that investors allocate capital or cash today to achieve a return on their investment in the future. As the future is uncertain, so is the inherent riskiness of the investment along with the expected return. Therefore, evaluating a company is subjective, and investors attempt to assess their investment decisions based on expectations about the future leading to financial assumptions about the future. It is important to note that there is always a "buyer" and a "seller" in a transaction, where there is a tendency rather to focus on why investors are selling than why another investor sees of potential in a company, and therefore is buying the shares of the company.

The original version of this chapter was revised: Figure 2.2 has been updated. The correction to this chapter is available at https://doi.org/10.1007/978-3-031-05800-4_45

P. Lykkesfeldt and L. L. Kjaergaard, *Investor Relations and ESG Reporting in a Regulatory Perspective*, https://doi.org/10.1007/978-3-031-05800-4_2

A fundamental investor is not concerned with how another investor will price the investment and generally will not invest in a company for the sole speculative purpose of hoping another investor will pay more for the investment later. On the other hand, a speculative investor will price a company's shares independent of the business and the real economy.

Fundamentally, the more disparities there are in the aggregated assumptions of the investors, the more shares of companies are traded, and the more stability of assumptions, the fewer shares are traded. Thus, when major news about the economic situation, or about a particular company, is published or revealed, high price volatility instantaneously occurs in the relevant companies' share price since the news alters the market's aggregated assumptions about the future. This continues until the aggregated buyers and sellers reach an agreed price level where they perceive the share price reflects the news released in the market. Therefore, the market price itself is an aggregated perception or opinion of the market participants based on all information publicly available in the market.

Often quoted in the works of Robert J. Schiller,[1] Samuelson proclaims that the market is micro-efficient but macro-inefficient. Essentially, the prices of individual stocks are moved by new information from factors such as dividends, an announcement of financial results, macroeconomic changes and other types of material information. According to Schiller, this is opposed to the aggregate financial markets dominated by speculative bubbles.

Stock prices seem to reflect their intrinsic value in the long term more accurately. Therefore, a large and transparent flow of information available for investors today related to fundamental valuation can offset financial markets "noise" from speculative bubbles. A large and transparent flow of information is subjective, and therefore, it is in the company's best interest to be as transparent as possible to avoid speculative bubbles. Schiller[2] sees bubbles as a combination of irrational social psychology and imperfect information, which results in:

> A situation in which price increases spurs investor enthusiasm which spread psychological contagion from person to person, in the process amplifying stories that might justify the price increase and bringing in a larger and larger class of investors, who, despite doubts about the real value of the investment, are drawn to it partly through envy of others' successes and partly through a gambler's excitement.

Factors that Determine Stock Prices

All stocks have a daily volatility level, which is irrelevant for IR, as it cannot be tampered with. A relatively higher volatility is typically the result of an increased uncertainty in the financial market about a company's future. However, best-practice IR can provide significant value to a company in terms of added market capitalisation in the event that a company's volatility and risk premium is reduced. Later in the book, we explore IR tools and methods to provide value as a best-practice IR. However, these considerations are meaningless if IR is not knowledgeable of the factors determining stock prices.

The investor relations officer (IRO) must clearly understand the factors that investors focus on and clear insights to address questions, ideas, and risks that investor acquire insights about. The best IRO who understands these factors that most of the company's stakeholders look at. It is complicated, if not impossible, to determine the exact factors that impact a particular company and weigh their importance precisely, which is also reflected in constant price moves of stock prices despite little to no new information in the market.

Stock prices change on the release of new information from various factors and are the aggregated assumptions of the investors to determine the cash flow of a company. However, some factors are more important than others regarding the individual company. Therefore, the company should be the market participant with the best understanding of the factors.

One method to control the most important factors in applying the marketing tool founded by F. Aguilar,[3] is the PEST model (abbreviation of political, economic, socio-cultural and technological factors). With constant analysis of these factors, a company can better understand its status while receiving insights into how the company should conduct its long-term planning, areas of threats, and insights into opportunities. The PEST model has later been expanded into PESTEL-DC, including environment (and ethics), legal, demographic and competition (Fig. 2.1).

Another method is Porter's five forces model,[4] where the competitive situation of a company or market is explained by the bargaining power of suppliers and customers in a market where there is a threat from new entrants and substitute products or services in the market. It is also important to split these groups into subgroups; for example, customers can be split into a company's customers of businesses (B2B), consumers (B2C) or governments (B2G). These factors describe the rivalry dynamics among existing competitors, which are useful to estimate the future cash flow of a particular company (Fig. 2.2).

The factors from the PESTEL-DC framework and Porter's five forces model are all constantly moving macro factors that impact a company's ability to perform in their operating setting. In addition, these are factors that the

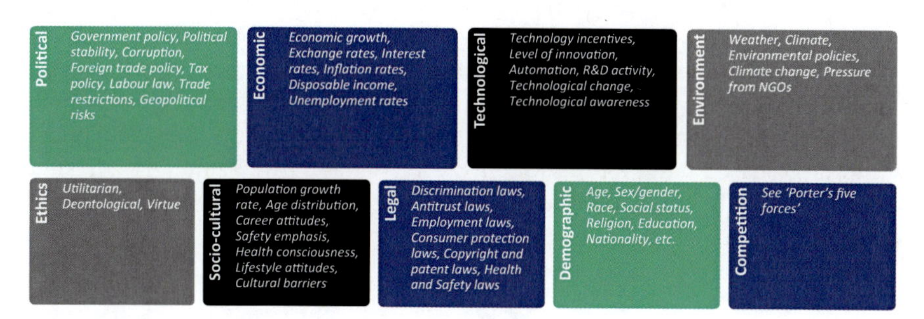

Fig. 2.1 PESTEL-DC model (*Source* Aguilar, F. [1967])

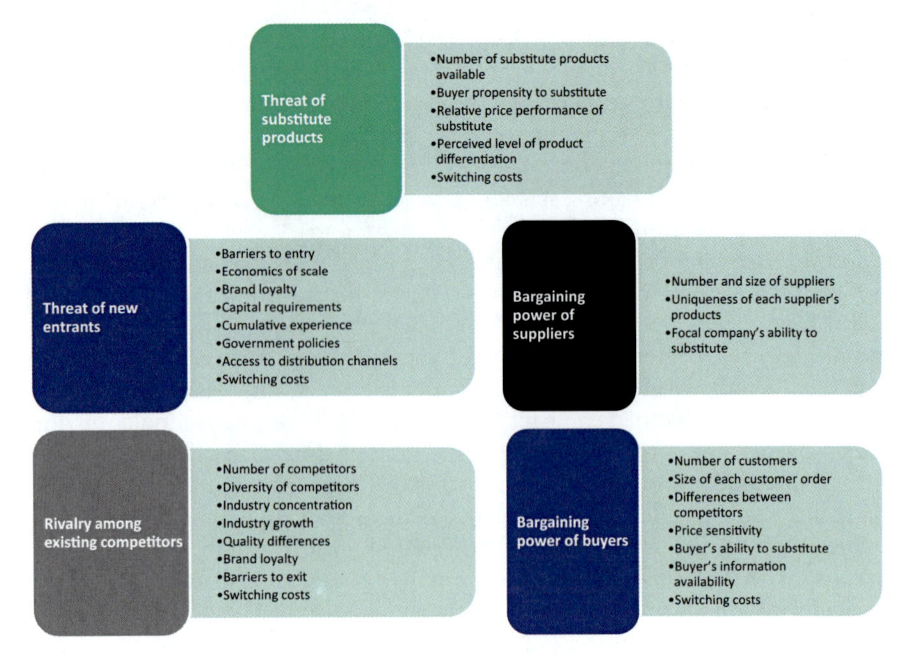

Fig. 2.2 Porter's five forces (*Source* Porter, M. [1979])

financial markets are attempting to price in and are therefore factors to be aware of when communicating with the financial markets.

Communication with Financial Markets Participants

To be clear, when communicating with investors, it is evident that company representatives may never break the law or cross clear regulatory lines by providing selective information to some investors at the expense of others. That is a complete given. There is a clear difference between providing granularity on market factors and scenario analysis of a company's considerations and possible actions to be taken, e.g. in connection with reporting, webcasts, etc.—versus providing selective market participants with sensitive non-disclosed material information.

The financial markets industry has historically and is today a very social setting. The market itself is the interaction between buyers and sellers, which has gone from physical to digital. However, the function of stock prices as the aggregated assumptions of investors remains the same. As a result, understanding the flow of information is essential to understand how to operate.

Information is king on the financial markets. As a result, the investor with the most information or highest access to information is better able to understand the market versus the aggregated assumptions of all investors (the market price). Therefore, information has an important role in the pricing of shares.

As markets are largely digital today, information travels faster and faster through the market participants and thus allows the market participants to price new information. However, information is also discussed, questioned and combined with more granularity gradually by professional investors. As a result, social interaction constantly occurs within the financial markets attempting to understand and debate the new information for its long-term implications.

The most important material information is distributed through company announcements and financial reporting through the stock exchanges or electronic wire, or provided at public webcasts, etc. However, other relevant information starts somewhere else. It travels through the market faster or slower, especially equity analysts (also referred to as sell-side analysts employed by brokers and investment banks; contrary to buy-side analysts employed by institutional investors) and the media act as efficient sources of spreading information as quickly as possible. In a sense, they act as constant detectives by attempting to uncover and distribute information that influences the aggregated assumptions of the market participants.

Types of Markets and Information

Successful long-term investing is a difficult discipline. The most difficult part is understanding the behaviour of the financial markets, which is based on the information accessible in the market. However, the crowd behaviour ultimately dominates decision-making and is responsible for the changes in sentiments and thus stock prices. The market participants attempt to piece together information by finding data points and ideas, discussing them with other market participants, and piecing together the information, trying to uncover an information advantage.

There are different types of traditions to view the financial markets, commonly considering market efficiency as strong, weak or semi-strong. For example, under the supervision of renowned economic scholars Merton Miller and Harry Roberts, Eugene Fama published a Ph.D. profound dissertation in 1964,[5] finding that stock prices were not very forecastable. As a result, it was difficult to predict the financial markets consistently, which later resulted in naming the efficient market hypothesis.

Strong, efficient markets are essential to perfect the financial markets and reflect all available information. With this assumption, investors cannot generate risk-adjusted excess returns relative to the market, not even with insider information, as all information is included in the existing price. This includes financial and non-financial information. On the other hand, semi-strong markets assume that publicly available information is not possible to generate excess returns (only insider information), and weak efficiency assumes that historical data cannot be used to predict future returns, but other information can.

Scholars have investigated anomalies that deviate from the strong, efficient market that can be explained with mispricing, unmeasured risk, arbitrage (deviation of two financial assets that should otherwise correlate), selection bias and behavioural and sentimental finance. These rely on that the efficiency of the market can be market can be semi-strong or even weak in certain scenarios. Examples can include sentiment in the aggregated investors willing to speculate, optimism and pessimism, and stocks with high growth potential, no cash flow or dividends with no tangible assets, giving investors little transparent information to value a stock. This leads to a high divergence between expectations and value.

The basis of attempting to piece together information to predict future cash inflow and outflow, i.e. net cash flows, for an extended period and therefore create an information advantage fundamentally assumes markets of weak efficiency. It is assumed that the financial market mistakes are corrected over time. An investor can also assume that the market correctly prices some stocks but misprices others relative to the correctly priced stock. Therefore, it sees the difference between them corrected over time.

Assuming weak market efficiency, most buy-side investors and equity analysts today attempt to apply a Mosaic theory or Scuttlebutt methodology to find information advantages. This is essentially combining public, non-public and non-material information about a company to determine the underlying value of the share in the future.

Legendary investor Phillip Fisher (1958) popularised the concept of the scuttlebutt methodology, where piecing together information from first-hand sources from discussions with expert networks such as employees, competitors, experts and analysts allows gaining an information advantage against the aggregated assumptions of the market about the future. It is said, we understand, that world-famous investor Warren Buffet was influenced by Fisher and Buffet's business partner, Charles "Charlie" Munger, to change their "deep-value" investment philosophy towards investing in high-quality companies applying this methodology.

Please note that despite their professional and theoretical applicability, the mosaic theory and the scuttlebutt methodology are often misused as a legal defence. This is typically in cases where investors have received and traded on non-public sensitive and material inside information, which is a strictly illegal practice that we naturally distance ourselves from.

Fig. 2.3 Risk premium of a share price

Risk Premium

Finding an information advantage can be extremely valuable as information is king in financial markets of weak efficiency. It is assumed that the information advantage is the difference between the intrinsic value of a company and the shares' market price. However, a company should also have a clear internal management view on its intrinsic value. As previously asserted, a company should be the market participant with the best understanding of the underlying factors, financial and non-financial, influencing its business.

Using appropriate risk factors, the chief financial officer (CFO) and the treasury function of a company may use all available information to collaborate with e.g. investment bankers or financial advisers to discover what they believe to be the true and fair intrinsic value that may be updated on an ongoing basis. As we already see signs where the risk premium is influenced by the company's ESG approach, it is important to consider these factors when determining a fair intrinsic value of a company.

As opposed to information advantage, from a company's perspective, the difference in intrinsic value and the market price is instead known as the risk premium. It is a consequence of uncertainties in the market. If a company realises a risk premium in connection with its shares, they essentially assume that investors are mispricing their shares. If the company uses realistic assumptions on their future performance and an appropriate discount factor to find the intrinsic value today, and it deviates from the market expectations—it is then in the company's interest to mitigate the risk premium with an increased focus on IR (Fig. 2.3).

Best practice IR is not to act as "used car salespeople" and "talking up the share price" but as professional information agents that assist in ensuring that market participants include fair assumptions in their valuations to price the shares following the intrinsic value gradually. Therefore, increasing knowledge of the company reduces uncertainty and the related risk premium so that

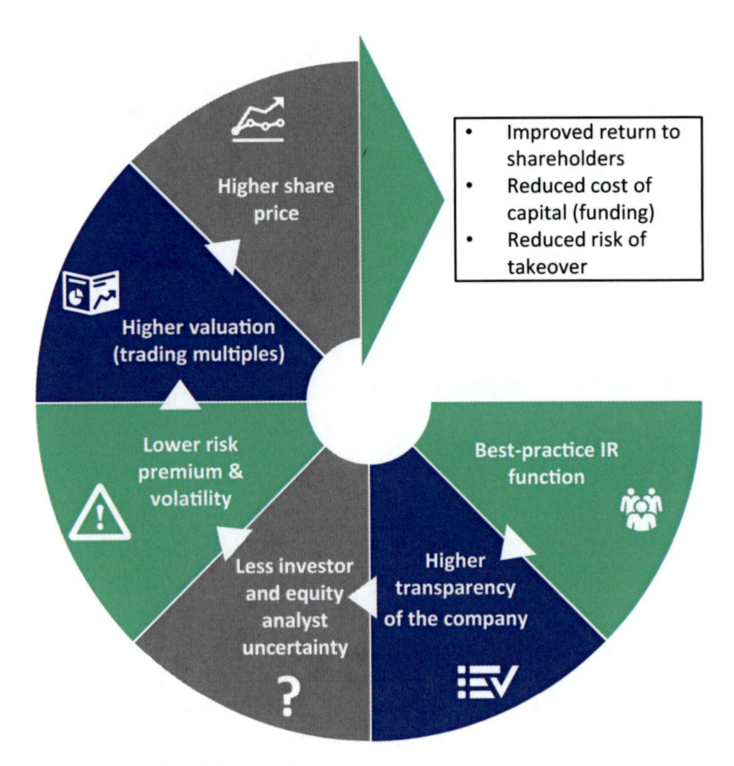

Fig. 2.4 Lowering the risk premium

the market price better reflects the intrinsic value. The company representatives (board of directors, senior management and IR) should always have a fair balanced view of the company discussing both positive and negative factors so that financial markets participants may obtain a balanced perception of the company (Fig. 2.4).

Some companies, including selected external IR consultants, view IR as a marketing function for the sole basis of inflating share prices. This is in some situations the case for companies planning to raise capital on the financial markets to expand their business, or to exit their business. Ultimately, this is not in the interest of all financial markets participants and will ultimately decrease the company's credibility, and hence increasing the risk premium. Eventually, such a company may find it more difficult to raise capital from its investors in the future (Fig. 2.5).

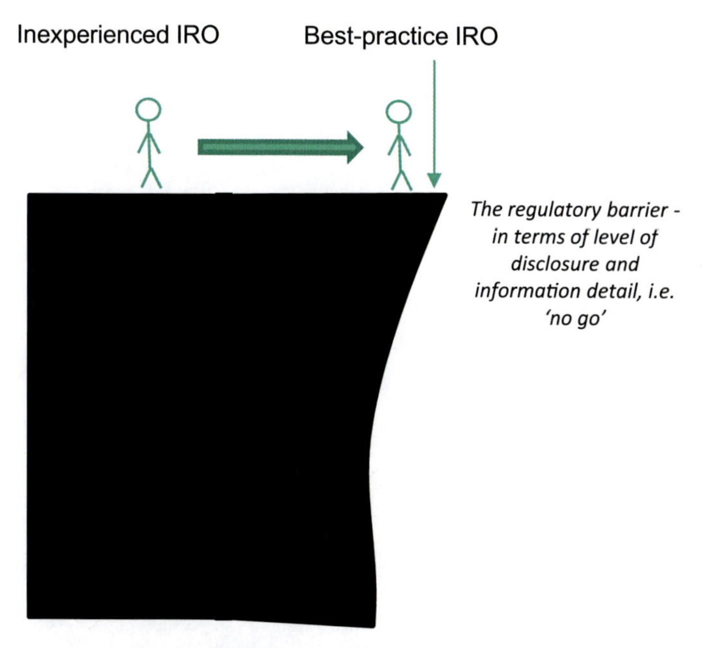

Fig. 2.5 Investor relations cliff—always be aware of the regulatory barrier

References and Sources

1. Jung, J., & Schiller, R. J. (2006). *Samuelson's dictum and the stock market*. Cowle's Foundation for Research in Economics, Yale University.
2. Schiller, R. J. (2013). *Speculative asset prices*. Price lecture, 8 December 2013.
3. Aguilar, F. (1967). *Scanning the business environment*. Macmillan.
4. Porter, M. (1979). How competitive forces shape strategy. *Harvard Business Review*.
5. Fama, E. (1965). The behavior of market prices. *The Journal of Business*, v. 38(1).

CHAPTER 3

Ethics in the Financial Markets: Why a Solid IR Framework Is Key

Acceptable and Unacceptable Information

To mitigate a risk premium, it is in the company's interest to provide the financial markets with relevant and transparent information on factors that allow the financial markets to estimate an intrinsic value which is more in-line with the company's view, and thus price the company's shares accordingly. However, it is essential for company representatives to distinguish between information that is acceptable and non-acceptable to disclose (please see the "Standards of Practice for Investor Relations" by the NIRI (National Investor Relations Institute)[1] for more details).

A company may provide granularity and deep-dive thoughts into the competitive markets and explore the factors that determine the share price. However, the company may not selectively share price sensitive, non-public information with the market participants. Such information includes information that a reasonable investor would consider important in making an investment decision.

Therefore, when communicating with stakeholders, most of the quantitative information provided to investors should be the information provided in the company's formal financial statements and announcements. It is best practice for other quantitative information to be in "ball-park" numbers, which may be shared on public webcasts. It is also vital for a company not to share information that could hurt its competitive situation.

A company should rely more heavily on qualitative information relating to the factors determining the market. For example, a company that may not yet have given their financial guidance to the following financial year may still be asked into sales growth expectations. Therefore, instead of providing actual estimated quantitative numbers, e.g. 8–10%, a company may point

P. Lykkesfeldt and L. L. Kjaergaard, *Investor Relations and ESG Reporting in a Regulatory Perspective*, https://doi.org/10.1007/978-3-031-05800-4_3

towards consensus estimates (i.e. average expectations of the equity analysts following the company's shares) of independent external sources, naturally without endorsing these numbers, and on a supplementary basis provide qualitative feedback on how various factors, for example, competition and other factors, may impact these levels and an indication of how the company is positioned in various scenarios. In such a scenario, the company shares its qualitative thoughts, which is acceptable as well as general market practice when done in the correct manner.

Hence, if a company or any of its peers have equity analysts following its shares, a company can then point towards the relevant consensus expectations, and thereby give granularity on scenarios of why these estimates could turn out more positive, negative or on the level with the consensus. The more experience an IRO has, typically the better the above is exercised in terms of expectations management.

Guidelines on Acceptable Information and Expectations Management

To educate the financial markets, it is acceptable to provide qualitative information on industry dynamics, products, markets, pricing, competition, buying patterns and other factors. Transparent management and IR providing acceptable information and expectations management are considered trustworthy by market participants and can therefore reduce the company's risk premium by providing relevant information about the company, market, supply chain and industry.

In addition, and on an occasional basis, it is acceptable within current legislation to provide a deep dive on certain industry topics to provide the market with a clearer picture of the company. This allows the market a better ability to project the company's financials without the company providing sensitive quantitative information which may exercise inappropriate information that may influence the share price of the company. The art of communicating with the financial markets' stakeholders, including exercising relevant acceptable expectations management, is to gradually educate investors and analysts without having an instant or explicit influence on the formation of the share price because. On the other hand, if an instant share price movement may take place, unacceptable information has likely been shared, creating an unfair information advantage for some market participants.

Guidelines on Unacceptable Information

If market participants have a conversation with senior management or the IRO and are provided price-sensitive information that deviate from market expectations, such shared information is considered unacceptable, illegal (in all major financial markets) and sensitive material information. Typically, the company representatives may have shared too much quantitative information that deviates from market expectations about aspects of which the company itself has strong control.

An example of unacceptable information could be the company's intention to increase prices, reduce costs or decrease capital expenditure levels, which are three fundamental factors that increase the company's net cash flow and thus valuation. Independent of the market, these are the factors that a company can control. If such an event would take place, the local financial authorities should be informed immediately, and under their advice, it may be recommended to issue an announcement to all market participants containing the relevant information. Naturally, the company is urged simultaneously to seek legal advice.

Financial markets representatives know that a company's competitors also track the company's information flow. Therefore, the financial markets representatives are highly respectful that specific data points or qualitative information is not necessarily unacceptable information but can be sensitive relating to competition dynamics. A company should never share unacceptable information and likewise never share sensitive competitor information, simply to satisfy investors. If felt necessary, a company may use this as an excuse not to answer certain questions from investors.

Reference and Source

1. Niri. (2016). *Standards of practice for investor relations*. Disclosure.

CHAPTER 4

Understanding Valuation Methodology of the Financial Markets

Valuation Methods of Different Investors

As the company seeks to diversify its investor base, it is essential to realise the different approaches applied by the investors. We have earlier established that information is king, and share prices are moved by new information. The information itself is then included in various types of methodology.

The type of investor largely dictates the different methods. Established investors either assume efficient markets to be weak or semi-strong. The semi-strong oriented investor sees some share as priced correctly and others incorrectly relative to them. The investor considering weak efficient markets attempts to find fundamentally mispriced shares. The first can be defined as "top-down" and the latter as "bottom-up" investors.

The company also dictates the method, as a bottom-up valuation method of discounting future cash flows (DCF) at today's point in time requires stable and positive future cash flows to arrive at a realistic estimate. This is mainly relevant for so-called value companies, whereas growth companies focus on growing their top-line and market share penetration, with less focus on earnings and cash flow.

The Top-Down Perspective

Top-down approach investors tend to identify overall macro-trends and factors, and then compare companies relative to each other. Therefore, an investor will compare a potential investment with a similar company (or a series of them) known as "peers". Besides identifying relevant company valuation metrics ("multiples"), a top-down investor will also compare financial metricise such as growth rates and profitability. These can include enterprise value

P. Lykkesfeldt and L. L. Kjaergaard, *Investor Relations and ESG Reporting in a Regulatory Perspective*, https://doi.org/10.1007/978-3-031-05800-4_4

(EV) versus sales, earnings (at different profit and loss (P&L) statements) or a combination of growth rates and earnings margins.

With these tools, the investor will identify if the potential investment should trade in line with peers at a discount, on par, or at premium. The top-down perspective can also be applied using transaction multiples, which similar companies have been bought for. In the case of the latter, this will include a "premium of control". Therefore, applying these multiples to the potential investment signals a trading value of the company.

The key advantage of this investment approach, i.e. primarily using trading multiples, is that it is simple, fast and easy to apply. In addition, it is relatively objective and can be used as a supplement to other valuation methods. As essentially any multiple can be constructed.

The disadvantage is that the investor assumes that the peers are priced correctly, whereas the potential investment is mispriced. In addition, the process is somewhat simplistic and very few companies have exact or close peers' comparisons. Finally, the multiples themselves are already dependent on expectations, historical trends, different market exposures and risk premiums. Therefore, an investor tends to price an asset relative to how other investors would price it—rather than to its intrinsic value.

The Bottom-Up Perspective

Unlike the top-down approach investor, a bottom-up approach investor assumes completely imperfect markets, and that the market fundamentally misprices some companies despite the high accessibility of information. However, with a detailed insight into a company, an investor seeks to discover potential value in the company which is not already priced into the share price by the market.

The most common bottom-up method is the DCF approach of discounting the estimated future cash inflow and outflow by an appropriate discount rate. After having mapped a company's historical financials, an investor will subsequently forecast the future fundamentals in several stages, typically stage one (typically three years), stage two (up ten years) and stage three, a long-term terminal value—and hereafter aggregate the cash flows and then discount them back today to estimate the intrinsic value as per today.

The main limitation of the DCF method is that it includes many assumptions. The future is uncertain; however, the method assumes that visibility is relatively higher (albeit not high) in the initial three estimated forecast years, lower in subsequent four to ten years and very low beyond that. In addition, positive and stable cash flow in the forecasted timeframe is necessary. In addition, it is a resource and time-intensive exercise to study target investments to such a high degree.

In most companies, most of the value exists in the terminal value (as it has the longest tenure); however, from the average bottom-up investor's perspective, most relevant news flow impacts the company in stage 1 and secondary

Profit and loss (P&L) statement:

- Revenue: Assumptions on volume and price development
- Production costs: Assumptions on efficiency and inflation
- Operating costs (OPEX): Assumptions on efficiency and inflation
- Depreciations of assets: Assumptions on historical capital expenditures
- Financial items: Assumptions on interest rates, the company's capital and risk profile
- Tax rate: Assumptions on regulation and geographical sales

Balance sheet (BS):

- Receivables: Assumptions on efficiency and customer pressure
- Payables: Assumptions on efficiency and supplier pressure
- Inventory: Assumptions on efficiency and customer dynamics
- Consumer advancement: Assumptions on efficiency and customer dynamics

Cash flow statement (CF):

- Capital expenditures (CAPEX): Assumptions on capacity constraints in volumes
- Non-cash and IFRS16 adjustments: Assumptions based on a mix of fundamental and regulatory forecast

Fundamental assumptions of:

- Cost of equity (beta value): Assumptions on market volatility of the stock relative to the market
- Cost of debt (weighted average cost of capital (WACC)): Assumptions on the risk of the perceived credit risk of the company

Fig. 4.1 Key input variables in a discount free cash flow

in stage 2. Therefore, an investor can include more tangible assumptions to estimates in these stages, which can be amended as the company or external sources disclose new information.

In theory, company valuation is a science. However, in practical terms, based on our experience, company valuation is more of an art (combining the science) as substantial market knowledge and experience is required to perform a trustworthy valuation of a company (Fig. 4.1).

Abbreviations of Financial Reporting

It is outside the scope of this book to seek providing a detailed insight into financial reporting. However, for the sake of good order, we would like to clarify a few terms, which are also important abbreviations to understand to evaluate ESG in the context of the next chapter (Chapter 5: Integrating ESG in Equity Research):

- Sales or revenue: A function of volumes sold multiplied by the average price.
- Operating expenses (OPEX) includes direct non-capitalised costs, e.g.:
  - Cost of goods sold (COGS), i.e. the production costs directly associated with producing a particular product or service.
  - Research and development (R&D).
  - Building renovation measures.
  - Short-term leases.
  - Maintenance and repair.
  - Other direct expenditures relating to the day-to-day servicing physical assets.
- Capital expenditures (CAPEX): Major long-term capital costs and investments in physical assets that cannot be directly deducted from the income for tax purposes.

CHAPTER 5

Integrating ESG in Equity Research

Embracing tangible ESG

Total assets in sustainable investments are worth almost 10% of global financial assets, or approximately 35 trillion euro (40 trillion US dollar).[1] First coined in the UN's Principles of Responsible investing report in 2006, sustainable investing integrates and includes ESG factors in the investment decision. The underlying themes behind ESG are not new, especially governance, which has been a main investor topic throughout the twenty-first century; however, the integration in a tangible framework is growing fast. We explore the origin of responsible investing in Chapter 6 (Fig. 5.1).

Today, there is no international commonly approved standard for preparing sustainability ratings and analysis. As a result, these are harder to derive and often at a poorer quality compared to traditional financial data. Given the often relatively poor standard of ESG-related data, the aggregated financial community seems not yet to have fully embraced tangibly sustainable investing altogether; however, this is currently changing at a very fast pace on an international basis. An investor who applies tangible ESG data has different methods and strategies to integrate these factors; however, for most investors, they are viewed alongside the traditional valuation methods that we have previously explored. As all factors that can explain a company's future cash flows should be included in an efficient market price, potential ESG risks and opportunities should also be formed from a theoretical perspective. Therefore, we argue that given the benefits of ESG in the equity analysis, it justifies a trading premium for the best-in-class companies with respect to ESG. In addition to lowering the risk premium, all things being equal, good ESG is also becoming a prerequisite and license to operate for the majority of listed companies (Fig. 5.2).

P. Lykkesfeldt and L. L. Kjaergaard, *Investor Relations and ESG Reporting in a Regulatory Perspective*, https://doi.org/10.1007/978-3-031-05800-4_5

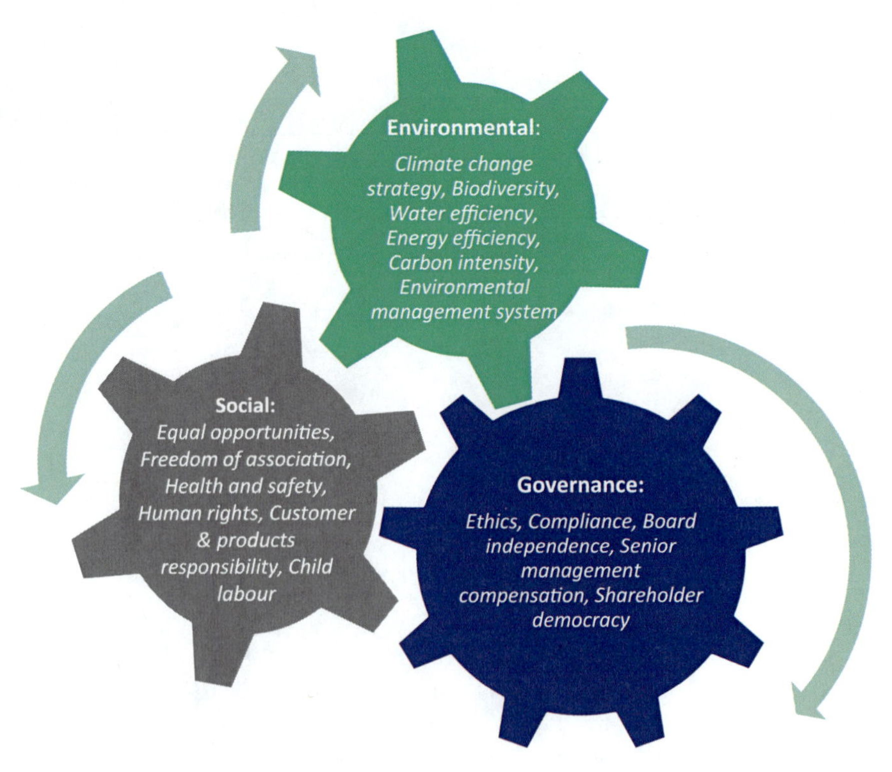

Fig. 5.1 Breakdown of ESG composition

As opposed to the neo-capitalist view that a company's main motive is to combine capital and labour resources to maximise a cash flow for its shareholders, the stakeholder theory allows stakeholder interests to be considered alongside cash flow. In the long term, value is created together with stakeholders and is thus a valid economic theory for the basis of ESG. The main motivations for an investor to include ESG in the investment analysis are:

- Allocate capital to solutions that are fundamentally better for the planet.
- A broader and deeper understanding of a company's value creation.
- Integrate academic evidence that ESG contributes to a better risk-adjusted return.
- Increase knowledge of potential downside risk factors.
- Allow more information to be included in the cash flow analysis.
- Embrace regulatory tailwind from the increased focus on sustainable investing.
- Attract capital from new and existing investors who embrace ESG.
- Engage companies through active ownership allowing co-creation of value.

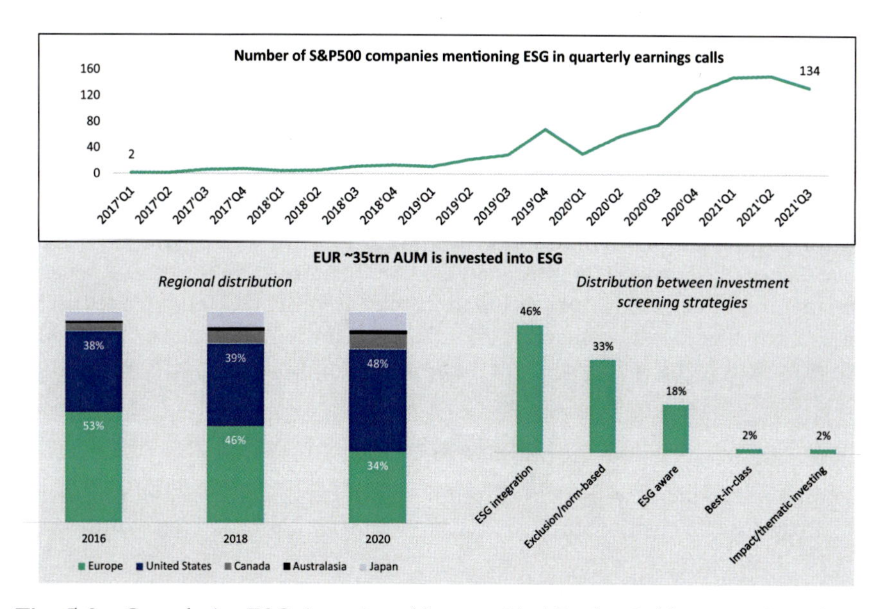

Fig. 5.2 Growth in ESG investing (*Sources* BlackRock, Goldman Sachs Global Investment Research, Factset)

Academic Literature on Best-in-Class ESG Companies

Including non-financial ESG factors into equity analysis is already being investigated from a theoretical perspective, which we believe is valuable insights for any board member, senior manager and IRO. Notably, research has been conducted on how best and worst ESG companies respond to the financial crisis, raising equity and generating profits.[2]

Academics have concluded that best-in-class ESG companies tend to be more conservative in its accounting and operating decisions, and that they provide more transparent financial information. In addition, they are less likely to engage in aggressive earnings management and less likely to violate legislation from stock exchanges and other regulatory bodies.

During the global financial crisis of 2008, best-in-class ESG companies tended to show higher profitability at the end of 2008 and in the beginning of 2009. Moreover, a higher profitability persisted during the post-crisis period. In general, these companies showed higher sales growth and higher gross margins around the 2008 financial crisis. In addition, they were able to raise more debt, and the economic effects were more modest.

The cost of equity is an essential input in the company's ability to generate long-term free cash flow. In general, best-in-class ESG companies display a lower cost of capital as they can attract more dedicated institutional investors and a higher degree of equity research analyst coverage. In addition, the study showed that companies moving from inferior ESG reporting to best-in-class ESG reporting measurably reduced their cost of equity capital. As a result, best-in-class ESG companies can typically raise considerably more new equity capital.

ESG in the Cash Flow Analysis

If a fundamental investor assumes markets of weak efficiency and applies a bottom-up valuation methodology of DCF, then there are several quantitative factors to include in the analysis. We believe ESG may have several fundamental effects on factors included in the PESTEL-DC framework and in Porter's five forces—and thus have an influence on the P&L, balance sheet and cash flow statement, along with the basic assumptions of the DCF modelling.

The Profit and Loss Statement (P&L)

A company can increase its sales through selling larger volumes and raising its prices towards its customers:

- B2B customers are increasing requirements of sustainability from their suppliers and may even be willing to pay higher costs if their supplier can offer a better ESG product or service versus other suppliers.
- Already increasing companies include ESG-orientated labelling (e.g. fair-trade, ecolabel, animal welfare, ecological, responsible sources, etc.) tailored towards B2C customers, allowing product differentiation to increase volumes but also sell at price premiums; it is therefore essential to document the supply chain associated with gaining these labels.
- B2G customers have increasingly made ESG reporting a prerequisite to bidding on large government contracts. In addition, they have begun to include performance bonuses (or penalties) based on CO_2 emission targets and accidents at work.
- On the other hand, an unfocused approach to ESG may lead to a worse perception. This can result in a company losing customers, volumes sold or are worse positioned in a price negotiation.

Companies can reduce their production costs by reducing their consumption of input variables such as energy, petrol, water, waste, pollution and production labour costs. For example, suppose a production company replaces a fuel source such as coke-fuel and coal with renewable energy. In that case, the company can, in some circumstances, reduce energy consumption cost of energy or even receive government grants supporting the energy cost likewise, if a company's transportation and logistics costs can benefit from electrification of the fleet. Due to a slow regulatory roll-out and perhaps expensive technology, sometimes companies may need to determine the trade-off of being a first-mover or protecting the company's profits in the shorter term. In such significant cases, the company needs to consider the non-financial elements of such a decision and communicate these to investors. If so, the financial market may conclude that the company's actions qualify for a reduced risk premium.

The most expensive component relating to labour is employment churn. If a company can facilitate a better workplace through more health and safety, better equipment and less pollution, and better education, this may increase motivation and decrease churn and related costs. In companies with high dependence on production labour, a professional investor or analyst will include assumptions on wage inflation, number of employees and churn, where a better workplace can reduce the latter. In addition, the well-being of employees also increases happiness, which in turn decreases churn, sick days and the risk of accidents.

With the inclusion of ESG, more administration generally is needed, and this is unlikely to be mitigated from lower usage of company cars, travel and expenses, which are good ESG factors. Therefore, transforming a business is always costly and including ESG strategy and administration will increase the company's operating costs in the short term. These are the relatively fixed costs that are not directly involved with the company's product or service—such as administration, offices, marketing, legal, accounting, typically increase. This is, however, natural and an essential element of responsible investing.

Depreciations are deducted in the free cash flow analysis to arrive at the operating profit. However, as depreciations are not classified as a cash flow, so it is re-added in the DCF. Depreciation is associated with the company's usage and tenure of equipment and therefore associated with the company's historical CAPEX on material and non-material assets. In the event that an element of CAPEX may focus on ESG solutions, it is vital to understand if this will impact depreciation in the future.

For example, the capacity and efficiency of electric powered batteries are improving significantly. Therefore, the depreciation rate of these assets may be different from standard petrol-powered equipment, where efficiency is slower. On the other hand, lower CAPEX due to more efficient equipment (versus costs) and higher subsidies will result in a lower depreciation base.

A company's financial items are mainly based on interest rates (the cost of borrowing); the company's capital structure; and the lenders' "perceived" riskiness. Suppose a company increases its focus on ESG. In that case, it may increase its internal governance functions and reduce the downside risk of default or regulatory scrutiny, while enhancing its collaboration with stakeholders and shareholder groups. By reducing the following risks, a company can decrease its financial costs:

- Credit (pre-settlement or settlement counterparty risks of default).
- Market (loss due to changes in market indicators).
- Operational (legal risks because of direct or indirect risk from inadequate or failed. Processes, people or systems from external events).
- Other risks (including reputation and strategic risks).

Lastly, businesses must pay tax in the jurisdictions where they operate and generate a profit. Good ESG practice is associated with transparency, and therefore, aggressive tax planning is viewed as unethical behaviour. We can easily assume that a higher level of tax transparency will lead to higher tax payments and is negative for free cash flow. However, we believe in the longer term that this will be compensated from a lower investment risk and a higher exposure from ESG-orientated institutional investors—which in turn reduces the risk premium on and increases the demand of the company's shares both leading to a higher share price.

The OECD and EU are continuously developing international taxation, and in 2019, the global sustainability standards board recommended that companies aspire for maximum tax transparency and disclose a country-by-country tax reporting schedule.[3] Further, in the light of high-profile tax avoidance cases such as the Panama and the Pandora papers, future legislation on tax transparency will only increase, and it will become more difficult for unethical companies to seek to mitigate its tax payments.

The Balance Sheet (BS)

Working capital is a key measurement of a company's operational efficiency and short-term financial health and, therefore, important in the DCF analysis. Working capital is the difference between a company's current assets and its current liabilities, and its ability to deploy cash efficiently. If cash is deployed less efficiently, the company must mitigate the cash outflow by raising capital through equity, debt, selling assets, increasing profits or growing at a slower rate.

To be specific, working capital is mainly affected by changes in inventories, receivables, payables and customer advances. Here, a company can reduce:

- Receivables by tightening its credit or collection policy from customers.
- Inventories with enhanced inventories or purchase planning towards its customers.
- Payables by negotiating extended payment periods with its suppliers.
- Customer advances by negotiating longer or even minimal customer advance options.

Therefore, working capital is primarily associated with a company's supply chain efficiency and its relationship with its suppliers and customers. We have little evidence to confirm a positive correlation between ESG and working capital; however, we are of the undocumented assumption that a strong focus on a sustainable business model and solid governance reduces the company's overall risk and should be positive attributes when negotiating with stakeholders.

The Cash Flow Statement (CF)

Following the company's earnings, depreciation, and working capital calculation, the DCF also requires assumptions regarding the cash outflow from CAPEX, outflow from acquisitions, inflow from disposals and inflow from dividends received from associates. For depreciation, we asserted that lower CAPEX due to more efficient equipment (versus costs) and higher subsidies will result in a lower depreciation base.

An investor will need to be informed on the types of projects and equipment the companies are purchasing, where ESG can be the main contribution and the types of businesses the company may purchase. In a mergers and acquisitions (M&A) situation, it is a valuable parameter to include ESG data about the target company and its financials, a topic we will revert to.

References and Sources

1. Diab., A., & Martin, G. (2021). *ESG assets may hit $53 trillion by 2025, a third of global Aum*. Bloomberg Intelligence.
2. Gianfrate., G. et al. (2018). *Cost of capital and sustainability: A literature review*. Erasmus platform for sustainable value creation.
3. Global reporting initiatives (GRI). (2017). *Tax 2019*.

CHAPTER 6

Valuation Methodology from the Perspective of Different Investor Types

Dedicated Available Resources Dictate the Methodology

Different types of investors apply several different valuation methods. Typically, the more established an investor is, the more dedicated resources, time, skills and efforts they apply to understand in detail the companies in their portfolio and target investments. The typical institutional investor is a more common operator of bottom-up investing, whereas the typical retail investor will apply the top-down methodology.

Bottom-up investing is more recognised by the academic and financial community, as Professor Dr. Pablo Fernandez emphasises: "The conceptually correct methods are those based on cash flow discounting. I briefly comment on other methods – even though they are conceptually incorrect – they continue to be used frequently".[1]

Bottom-up investing is time-consuming and assumes that the user can include information advantage dynamics in the modelling versus the aggregated market. This requires source input from companies, equity analysts and brokers, buy-side analysts, external and in-house experts and other stakeholders.

Top-down users can more simply view a company trading at a discount relative to its historic valuation or relative to a peer group and argue for an anomaly. This method is quite simple relative to a DCF and, therefore, particularly useful for filtering between shares or as a short-term valuation tool for some institutional investors, analysts and retail investors.

P. Lykkesfeldt and L. L. Kjaergaard, *Investor Relations and ESG Reporting in a Regulatory Perspective*, https://doi.org/10.1007/978-3-031-05800-4_6

Types of Investors

There are two main types of investor groups for equities, institutional investors and retail investors. Any company should strive for a balanced investor base to achieve the lowest possible risk premium because investors have different risk profiles, investment mandates and wish lists to determine which companies to invest in (Fig. 6.1).

Institutional Investors

In regulatory terms, institutional investors are classified as "professional investors" and include financial institutions such as pension funds, insurance companies, hedge funds, mutual funds (banks and non-owned banks), proprietary ("prop") trading at banks, family offices, foundations and university endowment funds. In simple terms, these investors act as custodians and invest on behalf of investors or other institutions with the primary goal of increasing the value of their investments through portfolio management of different equities.

Typically, financial institutions' investment arm (portfolio managers or PMs) attempt to build a portfolio of equities that deliver a high absolute return or a better relative return than a pre-specified benchmark. A successful track record is a reliable marketing tool to attract more assets under management from new or existing investors. In addition to a successful track record, professional investors also attempt to differentiate their offering towards investors. For example, proposing to innovate investment strategies; marketing an information advantage; well-performing portfolio managers; or other factors such as focusing on sustainability or active ownership.

As financial institutions have a high concentration of capital and a custodian status, they are highly regulated and in line with their investment mandate. Therefore, for a company, it is vital to understand their current and potential

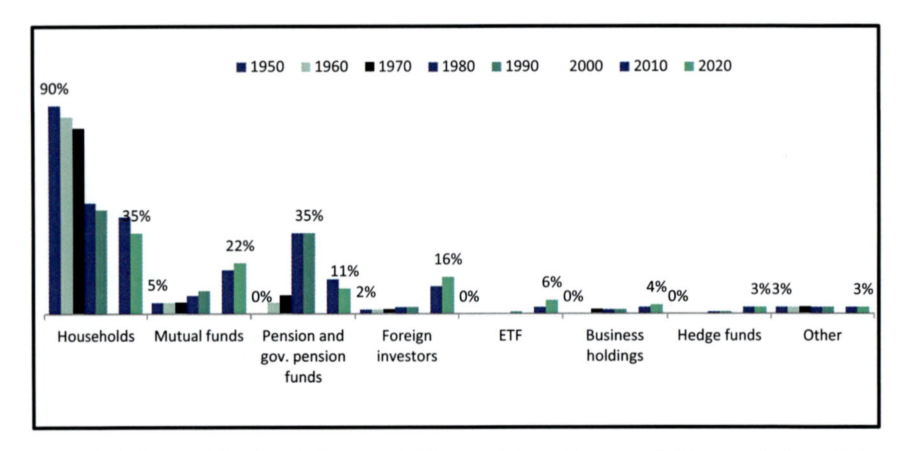

Fig. 6.1 Ownership breakdown of US equities (*Source* Goldman Sachs Global Investment Research)

inventors' investment mandates (which are often publicly available) to facilitate better their wishes in the company's IR-related communication towards them.

Institutional investors also include government institutions and companies; however, they typically have different motives than financial institutions. For example, if interest rates are low, an active government aiming to stimulate an economy can do so by allowing its central bank to buy assets (primarily bonds). The purpose is to act as a demand-side stimulus and put a floor under the market, which should lower societal risk (measured by the interest rates), increase consumer spending, and cut the cost of capital (especially borrowing of debt) for companies. Besides the Japanese government, equity investing among governments is relatively uncommon today; however, many governments invest in funds with an investment mandate orientated towards domestic companies to stimulate domestic companies.

Companies can distribute excess cash to investors, reinvest it into the company, simply keep the cash in the bank or invest the cash in the bond or share. We have already mentioned that a company can (often subject to a decision at a general assembly) buy back shares and thereby influence the number of shares available to investors to provide share liquidity. Yet, a company can also invest in other companies if they disclose so.

Retail Investors

On the other hand, retail investors (or private investors) are primarily classified as "non-professional investors" in regulatory terms. This is because they have a lower concentration of capital; however, they are much greater in numbers. Moreover, as they do not act as custodians of other investors but on behalf of their own household, they have much more flexibility to invest in different companies than a typical institutional investor. On an individual basis, they also have fewer due diligence requirements.

Historically, in particular in countries with a less developed financial market culture among retail investors (compared to the historically well-developed financial market culture in the US among retail investors), a number of companies have to some extent ignored the retail investor segment. However, with the constant global development over the past one to two decades of digitalisation, social media and financial market culture, it is now generally acknowledged among companies that the retail investor segment has become an increasing important and influential group of investors.

First and foremost, they typically offer a high proportion of daily liquidity in any given share. Secondly, they have become a significant force, e.g. via investor groupings through chat fora, etc., that at times may set the agenda vis-à-vis the companies, using the internet and the media to advance its views, and at times create virtual so-called shit storms which can be damaging among a variety of the company's other stakeholders, e.g. consumers, politicians, etc. Thirdly, retail investors typically support a larger mix of investors, and especially for small and locally orientated listed companies, retail investors may be a significant benefit supporting the financial structure of the company. On this background, more and more companies are now facilitating a much closer dialogue with their retail investor base, especially utilising the capabilities and benefits of social media.

Investment Strategies

Different investor types have different investment strategies. Institutional investors are dictated in the investment mandate stipulated in the public prospectus typically offered to their existing and potential investors. Retail investors are more flexible and can apply a combination of the strategies depending on the market opportunities that they find. We find that investment strategies are likely the most researched in finance; therefore, we will in the context of this book only highlight the most significant types below (Fig. 6.2).

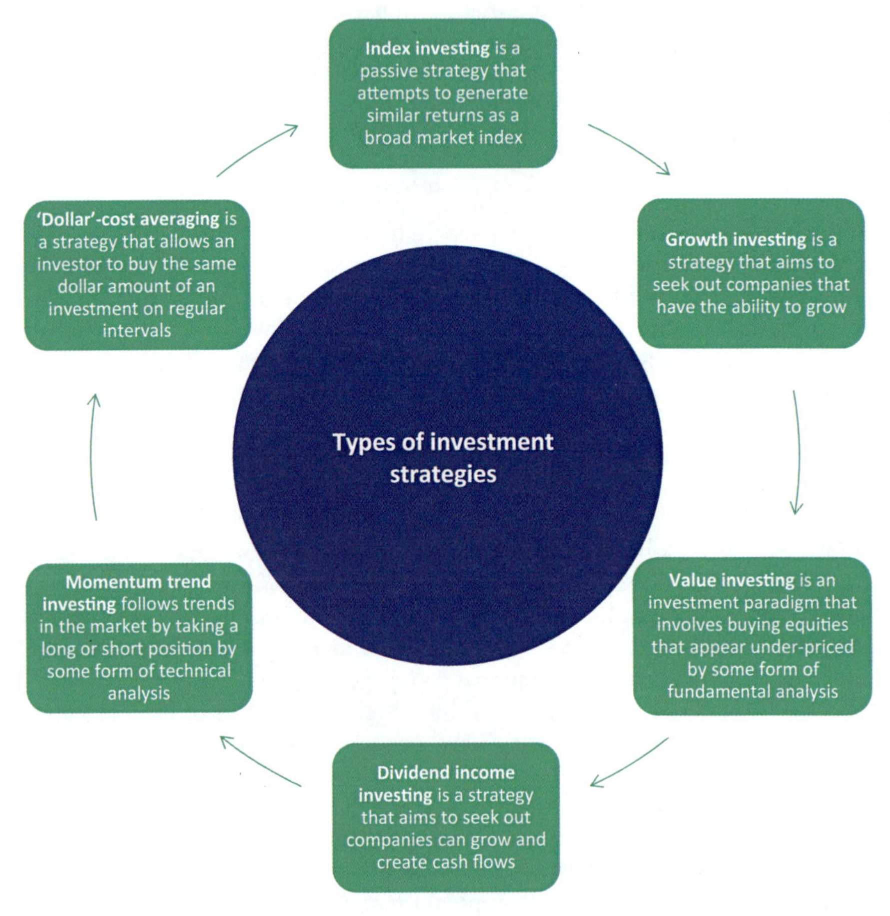

Fig. 6.2 Types of investment strategies

A summary of facts and best practices is illustrated on page 329.

Reference and Source

1. Fernandez, P. (2019). Company valuation models. *IESE research papers.*

PART II

The Participants of the Financial Markets

CHAPTER 7

Understanding the Financial Markets' Stakeholders and Their Motivation

You have to respect your audience. Without them, you're essentially standing alone, singing to yourself.

—Kathryn Dawn "K.D." Lang, Singer-songwriter and Activist

The Stakeholders on the Financial Markets

The financial community is comprised of investors, equity analysts, brokers and investment bankers. In addition to the market participants piecing together information by finding data points and ideas, they discuss them with other market participants to sense the market consensus and how to use new information to undercover an information advantage. As relying on and discussing information is essential, financial markets relations are socially structured. That means that the decision on the value of an asset requires social coordination and consensual validation. We have discussed that share prices are far more volatile than explained by movement in their intrinsic value; therefore, shares are priced according to fads and fashions among investors. This also explains the unpredictable nature of securities.

As a result, it is crucial to understand the actors, or stakeholders, of the financial community and what their fundamental roles and motivations are. Therefore, we explore the relevant actors in the financial markets—from banking institutions and different types of investors—to counterparts such as competitors, customers, IR consultants and the media. We outline their goals, motivation, daily responsibilities and how their interests are interconnected in the financial markets setting (Fig. 7.1).

P. Lykkesfeldt and L. L. Kjaergaard, *Investor Relations and ESG Reporting in a Regulatory Perspective*, https://doi.org/10.1007/978-3-031-05800-4_7

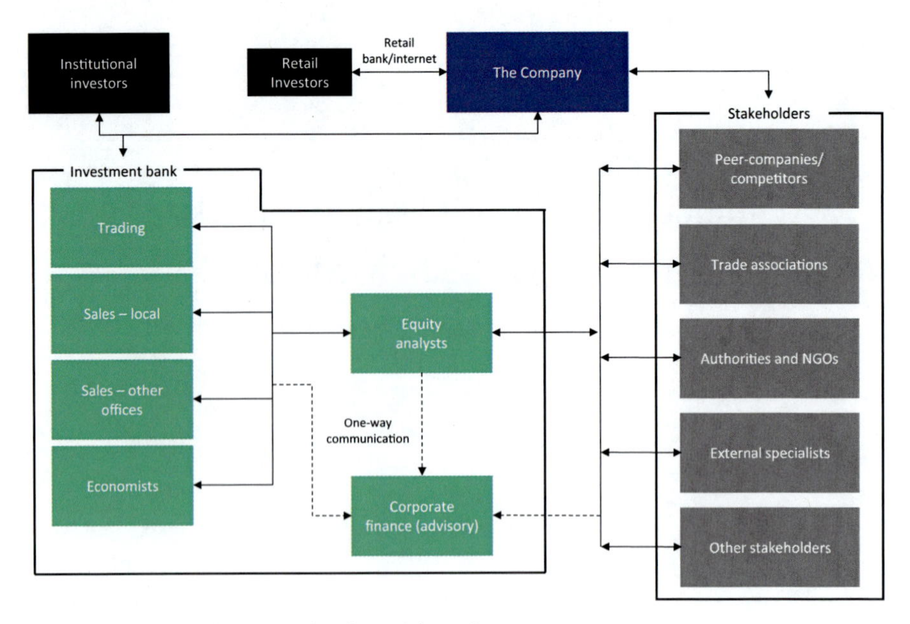

Fig. 7.1 Stakeholders on the financial markets

The Regulatory Separation

Investors are also referred to as shareholders (when they own shares in the company in question), portfolio managers and fund managers. The portfolio managers are employed in portfolio management companies, pension funds or investment associations. In addition, some of the most extensive portfolio management companies and investment associations have hired their independent equity analysts, also called buy-side equity analysts.

The key objectives of investors are portfolio and risk management within their trading space and mandates. However, for more sophisticated actors in the financial markets, the role of administration and marketing remains a significant part of the daily business and has increased in response to the regulatory wave.

Later, in Chapter 3, we discuss the essential regulatory themes for best-practice IR. However, it is initially necessary to recognise that the significant EU legalisation of MiFID II is a legal framework that aims to increase competition and investor protection, as well as levelling the playing field for investors. As a result, the legislation divides investors into three categories: eligible counterparties, professionals and retail. Investments banks and other financial institutions must have clear procedures in place to categorise clients and assess their suitability for investment advice and particular product:

- **Eligible counterparties** have the lowest investor protection as they are considered professional financial institutions, insurers, pension funds

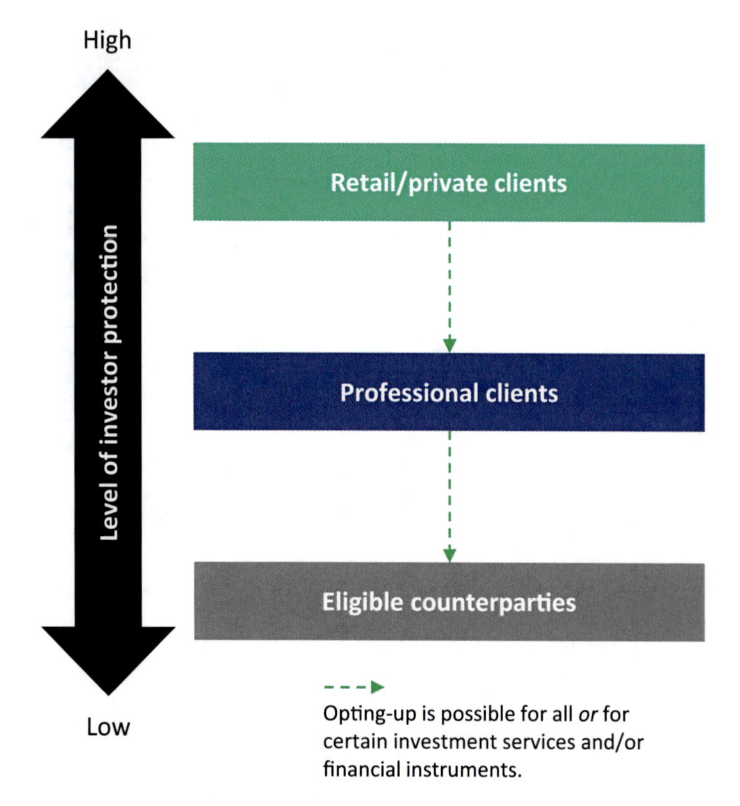

Fig. 7.2 Client classification under MiFID II

and governments. However, due to their established institutional nature and need to be highly regulated, it is assumed that they can work professionally.

- The category of **professional investors** is institutional investors and private individuals who possess the experience, knowledge and expertise to make their own investment decisions and duly assess the risk those investment decisions ensure. They must be regulated and meet several minimum size requirements and are also treated as institutional investors by investment banks.
- **Retail investors** are any clients not included in the other two categories. As a result, each investor must complete and confirm their knowledge in a suitability test before making investments in individual investment products (Fig. 7.2).

Investment Perspectives

A diverse investor base is valuable, and in the subsequent chapters, we deep dive into the main drivers and apparent differences between the types of institutional and retail investors. Despite a similar size of accumulated invested

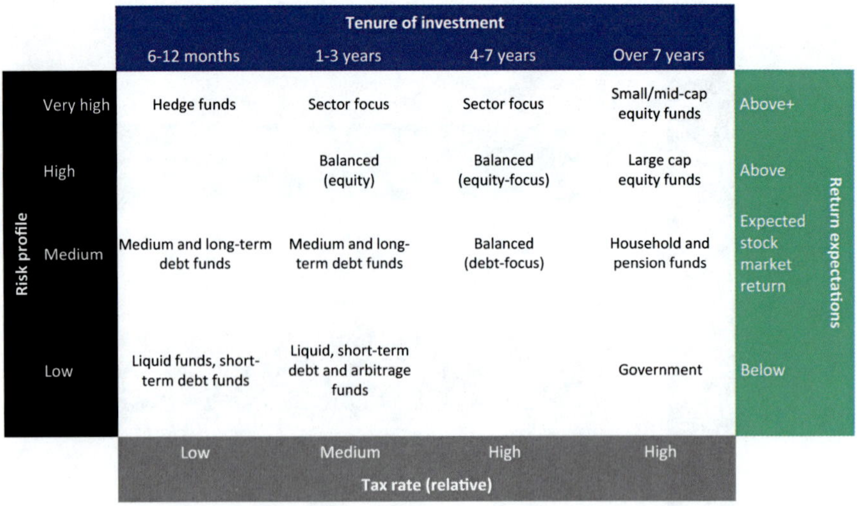

Risk profile	Tenure of investment: 6-12 months	1-3 years	4-7 years	Over 7 years	Return expectations
Very high	Hedge funds	Sector focus	Sector focus	Small/mid-cap equity funds	Above+
High		Balanced (equity)	Balanced (equity-focus)	Large cap equity funds	Above
Medium	Medium and long-term debt funds	Medium and long-term debt funds	Balanced (debt-focus)	Household and pension funds	Expected stock market return
Low	Liquid funds, short-term debt funds	Liquid, short-term debt and arbitrage funds		Government	Below
Tax rate (relative)	Low	Medium	High	High	

Fig. 7.3 Types of investment perspectives

capital in equity markets by institutional and retail investors, facilitating the two groups is quite different due to the concentration of capital.

The number of retail investors significantly outnumbers the number of institutional investors, and therefore, the IR function of a company cannot communicate individually to retail investors. However, it is possible to identify the different types of investors and the mix of interest among a company's shareholders (Fig. 7.3).

CHAPTER 8

Understanding the Role of Institutional Investors

Portfolio Managers (PMs)

In this chapter, our focus is on the buy-side as institutional investors. These are primarily professional institutions and individuals with significant capital under management. We discuss the role and motivation of portfolio managers (PMs) in tier 1 and tier 2 functions, buy-side equity analysts and other investor groups, including private equity and proprietary trading ("prop trading").

PMs are defined as professional investors who typically represent a particular investment fund with assets under management (AUM) backed by the significant means of capital handled by pension funds, banks and independent funds. The more prominent institutional investors (Tier 1s) also typically employ internal buy-side equity analysts who are independent researchers on behalf of their employer, contrary to equity analysts employed by investment banks and brokers who in principle and ultimately serve the interests of their own employers.

PMs use a combination of buy-side and equity analysts and their independent research with input from companies, industry experts and other specialists to make investment decisions. The mandates of PMs can typically be split into the following categories, which all include many sub-categories:

- **Long investors:** Investors who buy equities assuming that share prices will increase over time. These investors typically have a long-term investment horizon and need to sell equities to purchase others. Yet, over time they tend to re-balance their positions according to their risk tolerance and the price movement of their investments.

P. Lykkesfeldt and L. L. Kjaergaard, *Investor Relations and ESG Reporting in a Regulatory Perspective*, https://doi.org/10.1007/978-3-031-05800-4_8

- **Long/short investors:** Investors who buy equities assuming that share prices will increase, but these investors also have the option to short equities. The latter is the exercise of borrowing shares from a professional market participant and then selling them in the market—with a view to repurchase the shares at a cheaper price in the market at a later date, and then return them to the shares' lender, thereby gaining a profit per share corresponding the difference between the selling price and the repurchasing price of the shares. These investors attempt to balance good and bad investment decisions over time and use short positions to fund long positions.
- **Hedge funds:** Investors who use leverage (debt) to finance their equity positions (long and short) and are thus more speculative with a shorter time horizon. Hedge fund investors can take on more risks than the initial invested capital.

The fundamental goal of the investor is to seek the best possible combination of assets based on the investment mandate they have. A PM engages in investments, making capital available for the company that can best employ the capital at a given time. The PM seeks the highest return on the investments by making the capital available.

The primary motivation for the buy-side investor is the maximisation of accumulated returns from the investment positions included in the portfolio. Most shares are owned for accumulation instead of being a shorting instrument to speculate on decreasing share prices. The accumulated returns are measured either absolute or relative to a benchmark, where the benchmark is typically a selected stock index or an average of competing funds.

Good Performance is the Best Marketing Tool

In any function, higher performance results in an increased level of responsibility. Generally, suppose a PM can generate higher absolute returns than the relative benchmark. In that case, they can with a very high probability attract a higher capital inflow of capital and increase the assets under management. A PM is typically compensated with a base administration fee (typically 1–2%), however, with a higher fee potential in the variable incentive (usually a fee of 10–25% on the returns above specified threshold considerations). As a result, competition is high to achieve a variable incentive fee, which also results in a higher capital base for the administration fee; therefore, pressure has steadily increased on PMs to achieve a sustainable competitive long-term track record. The increased pressure has also been felt by companies as the sentiment in their share is typically significantly decreased if it fails to deliver on promises and expectations. In conclusion, typically PMs do not take lightly unprofessional behaviour, and unsubstantiated negative surprises, on behalf of companies in which they have invested.

The objective to achieve economies of scale on the invested capital, and to maintain a long and stable track record, has resulted in a highly consolidated industry. The relative number of PMs is therefore among the most scarce in the financial sector as it is difficult for new hires to be granted the responsibility of administering vast amounts of capital with no successful track record. Therefore, a limited number of PMs with a long and successful track record, and high competition between funds, have resulted in increased use of external managers and an uptick in "transfer fees" or guaranteed minimum bonuses in connection with recruiting PMs.

A Daily Target of Maintaining an Overview

PMs use a combination of equity analysts, buy-side analysts and industry experts to support their daily work. Other types of research and commentary include macroeconomic experts, asset allocation considerations (i.e., the weighting of different types of financial assets or equities at different times in an economic cycle) along broker commentary which is also included in their investment considerations. With these inputs, most of a PM's time is spent on:

- In-depth insights into the investment strategy, including the motivation of owning each of the current investment positions in the portfolio.
- Ensuring to have set up to proper information flow to track potential changes in the motivation to hold each investment position in the portfolio.
- Building and maintaining oversight of a watchlist containing potential new investment candidates to be added to the portfolio should the information flow changes.
- Maintaining an up-to-date view on news flow from domestic and international news sources.
- Reading annual and quarterly reports from companies that are owned; on the watchlist; peer companies (direct or indirect competitors); or e.g. customers, suppliers, etc.
- Engaging and sparring with in-house buy-side equity analysts.
- Filtering through the most relevant information, e.g. reports from equity analysts and brokers.
- Meet with the management or IRO of current and potential investment candidates along with peer companies.
- Participate in educational events hosted by investment banks and brokers (e.g. investor conferences), companies (e.g. Capital Market Days) and other industry experts (e.g. industry organisations).
- Conduct own analysis based on independently produced financial models or financial models supplied by equity analysts.

- Meetings and phone calls with brokers, equity and buy-side equity analysts in order to challenge assumptions and to compare own considerations to market consensus estimates.
- Sparring with the back-office trading department to ensure trading instructions are clear.
- Writing or supporting back-office staff in writing bylaws, prospectus and investor letters.
- Preparing investment recommendations for internal investment committees in the event of a proposed significant investment decisions.

As opposed to equity analysts and brokers who attempt to extract granularity or the possibility of hosting investor events on behalf of companies, PMs demand fair treatment in exchange for the capital they have invested in the company. PMs are, after all, part-owners of the company, and it is, therefore, crucial for the company—both selected senior management and the IRO—to create good and impeccable relations with the PMs and manage expectations. Consequently, it is vital for the company to include their principal shareholders in all communication provided to the market and make brokers and equity analysts aware of which investors are essential in hosting investor events.

Attempting to Find Information Advantages

PMs are aware that brokers and equity analysts provide all new relevant information directly to relevant investors, including that from the company. As competition is high among PMs, they have generally become more proactive in communicating directly with the companies that they have invested in. They will, therefore, regularly ask for calls before the company's quarterly silent period (pre-close calls), commentary of market news and other enquiries. Major institutional investors will also demand private one-to-one meetings with companies along with group meetings that equity analysts do not attend to get answers, or comments, on their considerations and assumptions.

While institutional investors are generally credited with a higher degree of overview, an increased level of detailed analysis has historically been credited to the equity analysts. Increasingly, however, PMs have today detailed analytical tools available and models that are on par with the ones of equity analysts. Therefore, the interaction between equity analysts and PMs has over the past years moved in the direction of challenging assumptions and considerations for best- and worst-case scenarios, rather than merely conveying news flow and recommendations to PMs. However, despite this trend, equity analysts still typically have the most daily contact with companies and more detailed insight into the companies they follow.

Types of Portfolio Managers

As outlined, a diversified investor base has many advantages for a listed company. Therefore, companies are recommended to distinguish between the types of investors, their size and the level of service required to satisfy their investment decisions. Major institutional investors (tier 1 clients) are highly consolidated investors with high AUM and require a high level of service and research (companies, macroeconomics, etc.) from investment banks and brokers. They are generally part of organisations (e.g. pension funds and investment arms of banks or investment banks) with high compliance requirements. In return for a high level of service, they deliver large trading volumes and research subscription payments to the investment banks and brokers.

PMs in tier 2 investment segment are generally more unconsolidated; they have less AUM and a less detailed view on investments. In addition, they are typical representatives of private high-net-worth individuals, investment arms of regional banks, family offices or investment start-ups. As a result, they tend to have a more flexible decision-making process with fewer compliance restrictions and, therefore, in general, have a higher opportunity to elevate risk.

As tier 1 investors tend to represent established institutional investors, they tend to have a strong reputation with high compliance requirements and detailed insights when making investment decisions. Therefore, a company will benefit from having tier 1 investors included in their shareholder structure as the inclusion of tier 1 in the shareholder base also serves as a credibility stamp towards tier 2 and retail investors (cf. Chapter 9). However, tier 2 and retail investors tend to provide more volatility and liquidity for the traded share volumes, which is a clear benefit for the investment decisions' tier 1 investors.

An Investment Universe: Investment Mandates and Themes

As established above, especially tier 1 investors have a highly robust methodology for their investment considerations—and their performance is typically measured against a defined benchmark. In addition, competition is high to attract capital inflow; therefore, diversification is needed to stand out from other investors. It makes little sense for most investors to focus on a broad universe of global equities without finding a niche or a focus area. Therefore, PMs focus on a particular mandate and invest based on specific themes.

Mandates can be narrowed to include specific industries, regions, sizes or types of companies—or a combination of them. Examples include SMEs (small-medium enterprises, typical companies with a market value between 100 million euro and 3 billion euro, although the definition may vary between geographical regions), EU-listed companies or perhaps a combination of both. It can also be companies that distribute high dividends to its shareholders or are classified as IT or capital goods.

In addition to the investment mandates, PMs also need to adhere to restrictions of the holding period; the minimum and maximum size of the positions in the portfolio; general allowance of short positions; and perhaps company-specific restrictions. These restrictions are pre-specified in a fund's (usually publicly available) bylaws and prospectus, as well as are followed up upon in its annual reports and investor letters. With a fixed mandate framework to navigate within, PMs can select, change and alter an investment strategy, as well as balance a portfolio of equities to match specific criteria.

Passive Investors

We have also found it relevant to discuss the role of exchange-traded funds (ETFs) from the perspective of IR, which has become an easy way to obtain a diversified investment portfolio at a low cost. Contrary to funds actively managed by institutional investors, these funds are passively managed, allowing institutional and retail investors to be exposed to specific themes. Therefore, the IRO needs to know which ETFs the company is included in, and which ETFs the company could potentially be included in.

The IRO cannot manage these relationships actively; however, by creating best-in-class financial and non-financial reporting, a company can be included in both multiple and a wide range of different types of ETFs. This is helpful as this increases the company's investor base, which ultimately supports the company's underlying share price and liquidity (Fig. 8.1).

Index funds (or ETFs) have become increasingly competitive over the past years due to better algorithms and increased speed of the trading systems. These passive investments have grown tremendously in the past two decades; however, they are primarily sector driven. Including specific factors and tracking with active management, integration can funnel the need for active approaches. In addition, technology-enabled mass customisation can assist in creating customised low-priced ETFs. Therefore, as it is impossible to manage ETFs today from an IR perspective. Further, we believe customisation development in the field can become highly relevant. It is, however, essential for the IRO to track the passive investment funds that the company is included in, as being either included or excluded from an ETF can have a strong impact on the company's share price.

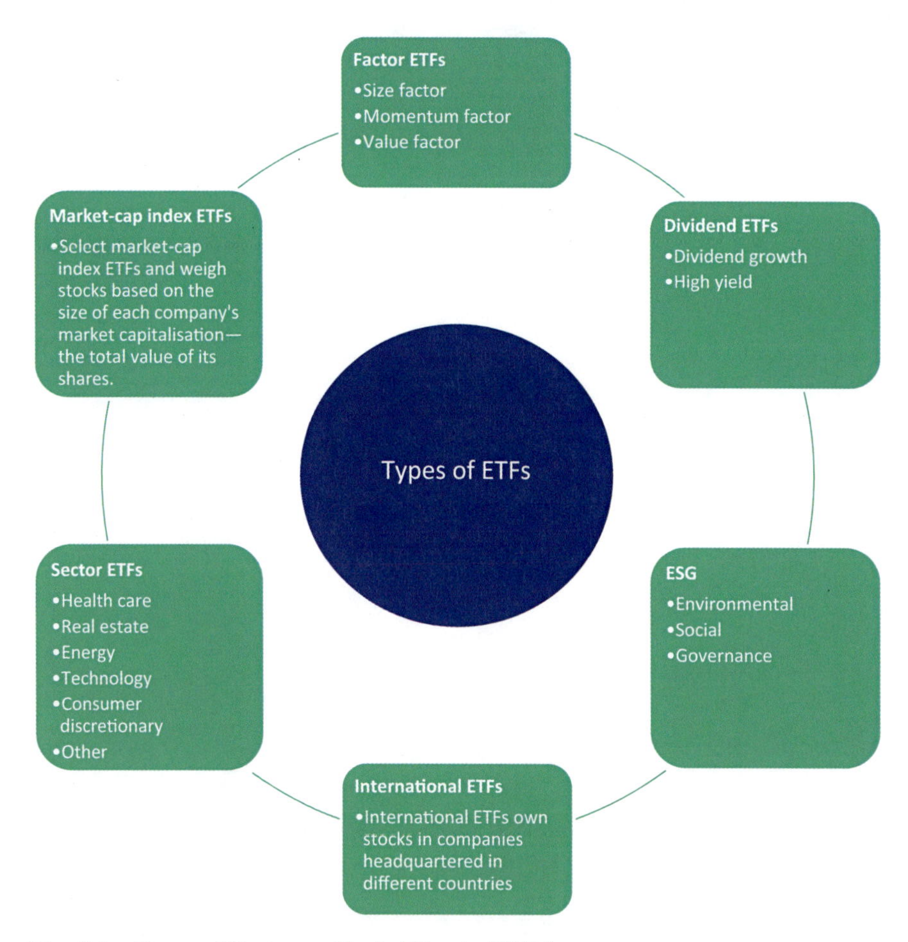

Fig. 8.1 Types of Exchange-Traded Funds (ETFs)

CHAPTER 9

How to Communicate with Retail Investors

Retail Investors

Retail investors are defined belonging to a different investor category than "professional" or "eligible counterparties" (ECP), and often trade on their accounts. Therefore, they must complete a questionnaire to determine which financial instruments are appropriate for the retail investor to invest in—before they may be serviced by a bank or a broker.

In the past, retail investors were not looked particularly favourably at by companies. Despite a relatively large share of equity ownership, strong market power and positive contribution to market liquidity, their decentralised nature made it difficult for companies to manage expectations and communicate with them. In addition, as they typically buy and sell shares more on news and sentiment rather than fundamental valuation, it means that they play a role in the determination of a company's risk premium or discount.

However, in the aftermath of the global financial crisis of 2008, and a higher distrust of the financial systems followed by lower interest rates, more retail investors have managed their own money. Combined with the surge of usage of social media, this led to a more robust investment culture among retail investors and, we believe, has become more sophisticated as an investment category. Retail investors on social media can share news, valuation assumptions and investment perspectives more efficiently, leading to the participants' ability to piece together information, and this way attempt to identify information advantages. This may be compared with the traditional relationship between institutional investors, equity analysts and brokers.

P. Lykkesfeldt and L. L. Kjaergaard, *Investor Relations and ESG Reporting in a Regulatory Perspective*, https://doi.org/10.1007/978-3-031-05800-4_9

Investment Banks Are Already Embracing Retail Investors

In addition, several professional investment banks have embraced retail investors and set up free-to-access digital corporate access and publishes freely distributed minor non-monetary (non-independent) research written by its equity analysts. This serves as a marketing platform for the investment bank and includes sponsored research paid by the companies themselves. Independent of investment banks, some financial companies also hire equity analysts for engaging with retail investors. A number of entrepreneurial banks, or trading platforms, have been founded in the past years, e.g. Robin Hood, Nordnet and Avanza, to cater for retail investors.

Retail investors as a group have become a professional organised force to be counted on and be aware of, just like societal consumer groups. Therefore, the companies' IR function needs to carefully consider its communication towards this investor group as it is the case with other investor groups.

Retail Investors in Connection with IPOs

During the preparation of an IPO (initial public offering) process, an investment bank (one or several) typically seeks to create interest from institutional investors and often attracts selected cornerstone investors who will commit capital to the process. Cornerstones are usually respected professional investment institutions, serving as an apparent "stamp of approval" in connection with an IPO, and thereby the company in question. In addition, it is assumed that the investment bank has conducted its own independent, in-depth due diligence on the company with the assistance of lawyers and accountants.

In collaboration with possible cornerstone investors, the company, the company's advisors and the investment bank then decide how many shares will be subscribed and at which price. These parties' interest is to conduct an IPO with a high "oversubscription" of shares, as this indicates the interest in the company at the pre-set price.

Having onboarded possible cornerstone investors and created a reasonable interest and commitment to the IPO, the investment bank then typically attempts to create interest among retail investors (although this may vary depending on both IPO in question and the geographical market). In addition to providing liquidity in the share, retail investors may also contribute to both obtain and maximise an oversubscription of the IPO. In the situation of an SME pursuing a larger IPO, it can be vital for the success of the IPO to obtain a significant retail investor engagement. In order to obtain interest among retail investors, the company and its advisers may use advertising (newspaper adds and marketing banners on social media and trading platforms) as well as host retail investor events via the local association of retail investors, trading platforms aimed at retail investors, etc.—all with a view to create a high level of interest among retail investors.

Once a company is listed on a stock exchange, a company with a high degree of retail investors in connection with the IPO will need to keep the retail investor community updated and engaged in order not to lose interest in the company and its equity story, and hence lose liquidity in the company's share.

In 2018, Spotify, however, disrupted the investment banking industry by conducting its own IPO process. In this connection, it is important to note that Spotify did not seek to raise additional money in connection with the IPO and was, and still is, a well-known household brand among consumers. In addition, Spotify was not concerned with managing institutional investors (corresponding to a free float of approximately 90%).[1] We believe this may serve as a textbook example of direct and digital communication with retail investors. However, the Spotify case can by no means be translated to the majority of IPOs and is extremely difficult to duplicate. Further, so far predominantly in the US, special purpose acquisition companies (SPACs) that typically raise capital without commercial operations at the time of the IPO have largely been aimed towards retail investors.

Communicating with Retail Investors

In recent years, as the private share culture has developed, many listed companies have focused more on the retail investor segment. This is partly due to the desire to exercise good shareholder communication; partly due to the recognition of retail investors' significance for the daily liquidity of the company's shares; and finally, partly due to the increased market influence of retail investors through social media.

The company's website is, to an even greater extent than for the institutional investor, the most important source of information regarding the company for the retail investor. This is because the retail investor typically does not have access to the extensive research produced by investment banks and brokers. Only the more prominent retail investors, e.g. high-net-worth-individuals, may count on receiving the equity analysts' research.

Any research received by retail investors will typically be research from medium-sized and larger investment banks, or commercial banks, or independent paid research companies, as most brokers do not serve retail investors.

A listed company's website also contains the company's annual report, quarterly reports, other stock exchange announcements, shareholder magazines, etc., which all serve as a source of information and inspiration for retail investors. Finally, companies typically also upload their investor presentations on their website that they have presented for institutional investors. Other sources for retail investors to obtain information about a company include:

- General meetings.
- Webcasts in connection with quarterly earnings reports.

- Shareholder meetings for retail investors.
- The media.
- Social media chatrooms.
- Digital IR (cf. Chapter 23).

Reference and Source

1. Pisani. B. (2018). Spotify's IPO disrupted Wall Street: What lies ahead now for unicorns looking to go public. *CNBC*.

CHAPTER 10

The Sell-Side/Equity Analysts, Brokers and Corporate Access

An Overview of How an Investment Bank Works

Inherently different to retail banking (or sometimes referred to as high street banks or commercial banks), investment banks only work with professional investors and companies. Investment banks are not in the business of providing a private individual with a mortgage loan, credit card or other types of banking activities typically associated with banks.

In general, an investment bank generates two sources of revenue defined as primary and secondary sources. Primary sources are fees collected related to corporate finance-led activities (based on advisory activities). A well-rounded corporate finance department typically operates within three segments: equity capital markets (ECM), debt capital markets (DCM) and mergers and acquisitions (M&A). It assists raising capital to companies from financial instruments such as issuing equity or bonds and provides advisory services.

With a high level of sensitive information in their possession, investment banks are highly regulated. Therefore, an investment bank must have strict compliance and confidentiality procedures in place. In addition, the investment bank has various virtual barriers of information flow (known as "Chinese walls") to block the exchange of information between departments, potentially resulting in conflicts of interest that could result in unethical or illegal business or other activities. For example, Chinese walls exists between the equity research departments and its equity analyst, sales and trading, and selected other departments within the investment bank on the one side, and the corporate finance department (inclusive of ECM-related activities) on the other side.

P. Lykkesfeldt and L. L. Kjaergaard, *Investor Relations and ESG Reporting in a Regulatory Perspective*, https://doi.org/10.1007/978-3-031-05800-4_10

Equity Capital Markets (ECM)

ECM include business activities, advisory and services related to listed or listed companies. The primary business is assisting a company in connection with an IPO, the process where a private company seeks to be listed on a public stock exchange. However, the company may, however, already be listed on a stock exchange of another country, thereby seeking a dual listing. In connection with an IPO, corporate finance collaborates with other functions of the investment bank, e.g. sales and trading (including corporate access), and the analyst from equity research. The collaboration between these functions and departments well-regulated both via the compliance department of the investment bank and via existing stock exchange and other current legislation which vary from country to country.

Once the financial instruments, e.g. ordinary shares, can be publicly traded, an investment bank subsequently generates secondary revenue in the sales and trading department from facilitating the trading of market participants. Secondary revenue is based on volumes traded ("flow") and the brokerage fee—the more clients trade in terms of market capitalisation (volumes times the share price), and the higher the brokerage fee they pay (in terms of percentage or rather a fraction of a percentage), the higher the secondary revenue.

In addition to secondary revenue, sales and trading are also involved with ECM activities. Some weeks before a company announces an IPO, known as "an intention to float" (ITF) announcement, sales and trading are educated by corporate finance and the equity analyst on the company. In collaboration with the corporate access team, sales and trading will then create interest among their clients and set up meetings with the equity analyst and the management of the company. The equity analyst and company management will then present their perspectives towards the investors on a roadshow.

Sales and trading will then collect and supply feedback and indicative orders from investors to the corporate finance department, who on this background will determine a valuation for the share offering in connection with the IPO in collaboration with the company, existing investors and possible institutional cornerstone investors that may have pre-committed to the IPO process. Finally, sales and trading will collect orders from investors, in a process known as "book building".

In general, investment banks employ equity analysts to produce research and valuations on publicly listed companies and recommend buying, selling, or holding them (alternatively: underweight, equal weight of overweight in terms of formal investment recommendations). Therefore, the equity analysts need to be independent of any conflicts of interest the investment bank may have. An equity analyst strives to provide respectable investment recommendations and ideas by conducting independent investment research, resulting in a resilient reputation, ideally—via various broker reviews—being top-rated in the given sector or industry.

Investment research from a reputable or top-rated equity analyst creates flow. In the past, the access to research, corporate access and trading fees were bundles together. However, due to regulatory changes in 2018 (MiFID II, explored in Chapter 3), the brokerage fees, research and corporate access had to be unbundled. This means that brokerage fees have since been dramatically reduced while research and corporate access are paid individually.

Concerning the primary revenue (for advisory services), the corporate finance department involves the equity analysts in an IPO process some months before sales and trading. The analyst will then write an independent research report to value the company in order to educate potential investors. In connection with an ITF announcement, the report is published and distributed to sales, trading and selected investors. The analyst will spend a few weeks on a roadshow, marketing the company and educating investors on the fundamentals of an investment case before the roadshow of the company management. Such a research report must not include an investment recommendation, nor a target price.

Equity Analysts

Equity analysts (we previously noted that they are also known as sell-side analysts) have varied areas of responsibility in an investment bank. Of the many functions, the relationship with investors is the most important, and these relationships are ultimately the greatest impact on the analyst's total salary and bonus. The most important stakeholder for an analyst is the internal trading and sales department.

The analyst's most important task in the longer term is to prepare differentiated, innovative research, i.e. unique analyses of the companies and industries that the analyst covers. In essence, the equity analyst's reputation is the main unique selling proposition, and this is what the analyst is rewarded for by investors. Of course, all equity analysts can make a simple follow-up on quarterly results. However, successful equity analysts work as constant detectives and uncover new relevant factors to determine the valuation of a company. An analyst, therefore, strives to provide information advantages for their clients (Fig. 10.1).

Equity analysts seek to create so-called innovative research. The analyst seeks to estimate the company's long-term sales and earnings development based on events that are difficult to predict. They also seek to estimate the companies' quarterly earnings. Internal and external prestige is associated with hitting close to the companies' later reported figures for sales and earnings. These quarterly earnings forecasts and subsequent follow-up of the realised financial figures are considered maintenance research. The preparation of these quarterly estimates takes a considerable amount of time, as there is a cash settlement here concerning the analyst's ability to estimate the company's accounting results correctly. This can best be compared to the companies' ability to hit their guidance.

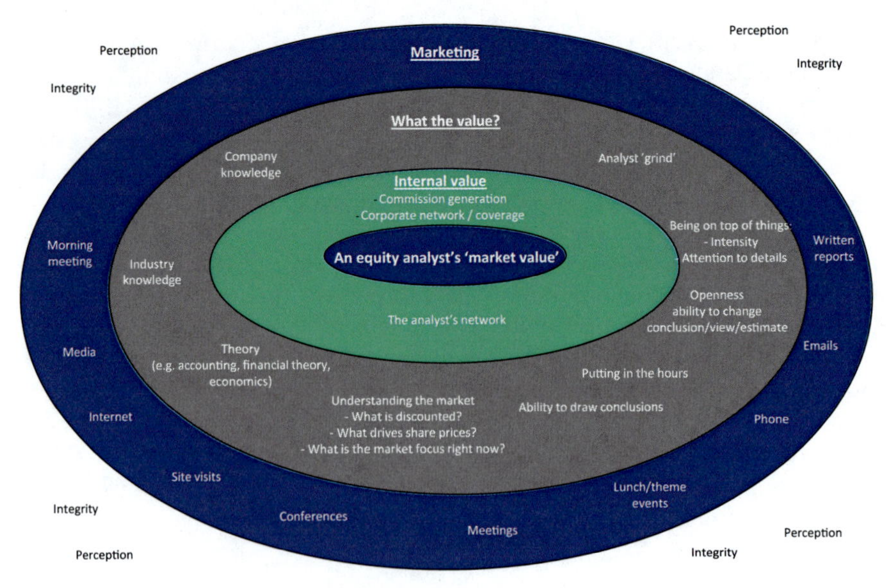

Fig. 10.1 An equity analyst's market value[1] (*Source* ABG Sundal Collier)

The importance of the accuracy of quarterly estimates is also because individual investors, especially the so-called hedge funds, which are very speculative, take short-term investment positions around the time of the companies' quarterly reports.

Finally, equity analysts are also concerned with the companies' capital structure and the influence of cash flow, share-buy backs and dividend payments.

The analyst works in many ways as the best business journalists in their research, though the analyst tends to go even deeper into the detail. However, two clear conditions distinguish these job types, as the analyst:

- Has a well-known smaller customer group, primarily the inventors.
- Has a key tool in their work—namely their financial model of the company.

The financial model created by the analyst is the core of the equity analyst's work. Through this, the analyst seeks to estimate the company's future financial development. The model is typically an Excel file, where company, competitors, markets and many other variables are modelled. The scope of the models varies. Some equity analysts have very comprehensive models; some have simpler ones. It also varies depending on the industry the company belongs to.

In other words, the analyst seeks to create as comprehensive an understanding of the company, the industry and the competitive situation as possible to forecast the future, including the company's future earnings. Then, depending on the analyst and the analysed company, they translate their estimates into either a bottom-up approach with a DCF or a top-down approach with relative valuation.

As a company's future cash flows and results are unknown, equity analysts merely present a view on how they believe a particular company should be valued. However, the assumptions and the weight associated with the different assumptions are subjective.

Types of Equity Analyst Research Reports

When equity analysts begin to analyse and publish regular research on a company (also known as initiating coverage of a company), they will typically start by registering all historical data from its P&L, BS and CF statements, typically in Microsoft Excel. In addition to the headline numbers, equity analysts will also register the breakdown notes of historical data (e.g. the top-line breakdown of business segments, extraordinary items, ESG data and types of depreciations and CAPEX). Then, based on market data and conservations with the company and experts, the equity analysts will forecast the company's development to conduct a bottom-up valuation.

When the comprehensive excel model is produced, the equity analysts will write reports about the company, its opportunities and risks and how to value it. There are different types of written reports, including:

- **Initiation of coverage:** A comprehensive report with a detailed description of the company and an introduction to market dynamics and valuation methods to value the company.
- **Preview:** An update on the analyst's expectations of an upcoming company updates and is a useful tool of maintenance research with minor estimate changes based on changes in market information.
- **Review/Post-result:** A update following a company's company update and, like a preview report, a useful tool to make estimate changes based on the company's updated trading update or quarterly report.
- **In-depth report:** Like an initiation report, a comprehensive report focusing on proprietary research shedding light on a specific theme that can change how to value the company. It is a matter of preparing share analyses that uncover conditions that have not previously been dealt with in-depth by competing equity analysts. These can be trends in demand, globally or locally, comparisons of competing products, assessments of next-generation products, conditions relating to research and development, etc.
- **Recommendation change:** a report has detailed the conditions of a recommendation change from a previous recommendation to a new, including buy/hold/sell, or underweight/equal weight/overweight.

- **Estimate changes:** Considering new data or an ad hoc update from the company or its peers, changes in the equity analysts 3-year estimates. This can be considering changes in the company's guidance, the announcement of M&A, CAPEX or similar.
- **Fast-comment report comment:** An unedited comment from the analyst published to all investors on a piece of specific news from the company. This is typically followed by an estimate changes report later to include the news into the equity analysts' estimates. A fast comment is also typically published following the release of a company's trading update to see how the released numbers align with the analyst and (and equity analysts' consensus) estimates. A good indication of a company's share reaction following new estimates is the earning deviation from consensus.
- **Sector reports:** A report concerning multiple companies within the same category. These examples include sector reports on banks, capital goods, construction or similar, where comparable market data influences multiple companies in a certain sector.

Communication Flow of Equity Analysts

The analyst is a specialist in a selected number of listed companies and one or more industries. The analyst's most important task is to ensure that their stakeholders and partners, including investors (customers), can make better and more informed trading and investment decisions. The three main areas of daily conversation partners that the analyst is in contact with can be divided into investors, colleagues in the investment bank and others (including the companies covered by the equities analyst)

In addition to mastering disciplines such as economics, finance, strategy and having an in-depth knowledge of the industry or industries that the analyst covers, it is essential for the analyst to have good communication skills. This is often where the top-rated analyst differentiates themselves. The analyst should also have a fine understanding of the financial markets and its psychology (Fig. 10.2).

Company Relations with the Equity Analysts

The company needs the analyst because:

- The analyst markets the company in their daily work, i.e. regularly discussing investment factors on its competitive dynamics. To keep the company on the investors' radar, it is in the company's interest for the analyst to constantly keep the investors updated on the company.
- The analyst is often helpful in arranging corporate access events (i.e. roadshows) domestically and abroad. This allows the company to expand its reach of new potential investors.

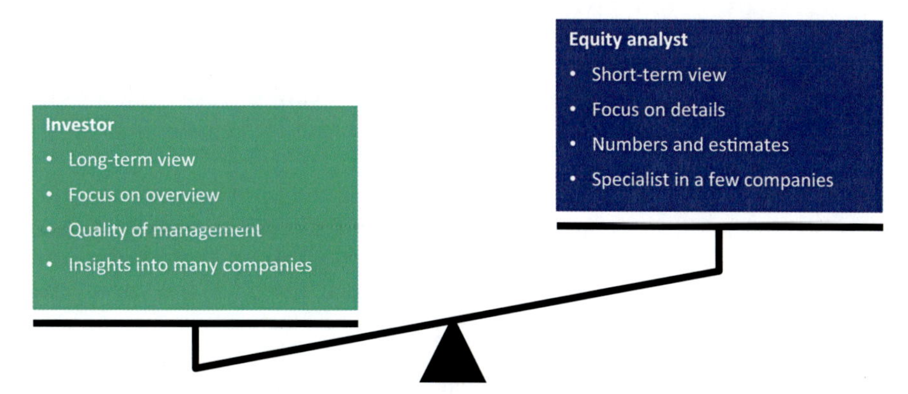

Fig. 10.2 Primary focus of investors and equity analysts

- A good relationship with the analyst typically has a positive contagious effect on the company's perception in the financial markets.
- The company often benefits from the equity analysts, as a rule, thorough market research and competitor information.

The analyst needs the company:

- An analyst will fall behind in servicing investors compared to competing equity analysts if they are not fully updated on the company.
- If an analyst falls behind on serving investors, it lowers the probability of allowing them to set up roadshows and other events.
- A bad relationship with the company entails reduced ratings by the investors and thus reduced secondary revenue for the investment banks and ultimately a reduced personal bonus.
- An analyst must be independent to build a reputation. There are examples that equity analysts may have lashed out with the company for some time due to some controversial views. Their views may prove to be correct. The analyst is typically rewarded for this by the investors. However, it is destructive for the analyst to have a strained relationship with the company in the long run.

Equity Analysts' Relationship with Investors

Investors are the equity analysts' customers and primary stakeholders. The investors' brokerage fees and research payments pay for the equity analysts' salaries. The analyst continuously seeks to keep investors updated both by telephone and via e-mails and its equity analysis regarding developments in:

- The analyst covers the shares.
- The industry in question that the analyst covers.
- Relevant subject areas of relevance to the company or industry in question.

Furthermore, in collaboration with sales, trading and corporate access, the equity analysts arrange events with expert speakers such as external specialists and lecturers and the companies covered by the analyst and peer companies. As the financial markets are a social setting, these events intend to shed new light on new and relevant investment factors that can ultimately help to improve the quality of investors' investment decisions.

The analyst, whose work capacity naturally has a limitation, typically has ongoing contact with the most critical 30–50 institutional investors who follow the companies covered by the analyst. The remaining institutional investors are kept updated by the brokers, also known as stockbrokers or the sales and trading department.

Brokers (Sales and Trading)

The broking side, i.e. the sales and trading departments, is the investment banks' trading link between investors and the listed equities, or shares. Their main revenue stream stems from trading orders and the price difference ("spread") between the buy and sell price associated with executing the orders. The broker's primary task is to keep investors updated on the financial markets' movements, including what is happening concerning individual companies. This typically takes place via a morning call with the investor and then continuously throughout the day.

Brokers act as customer managers and seek to ensure that investors are satisfied with the overall services, they receive from the broker in question, including the equity analysts. For brokers to perform their work satisfactorily, a daily internal morning meeting is generally held, in which equity analysts, sales and trading participate, and the ECM of investment banks attend. The equity analysts present their relevant investment cases, ideas and news, while sales and trading may ask questions.

Most brokers have a local morning meeting between 07:00 and 07:50 local time, and an international meeting between 07:50 and 08:10 local time, containing the most important news from equity analysts or potential ECM business. Other offices of the investment bank will join these morning meetings. The role of the analyst includes keeping sales and trading well informed about their shares throughout the day. Consistently, the analyst needs to react quickly and accurately, and interpret all significant news correctly to all their stakeholders. The traders are yet another of the analyst's stakeholders.

The responsibility of the traders is to:

- Trade shares with traders from other domestic and foreign brokers, i.e. a kind of wholesale market for the shares. The traders are not investors but intermediaries who ensure that the investment bank acquires or disposes off a given share at the most advantageous price at the time in question. Traders do typically not have any direct contact with the investors, i.e. the customers of the broker.
- Manage the investment bank's own holding of shares. When an investment bank's brokerage arm, also referred to as the broker, acts as a market maker in each share, the broker generally needs to have an inventory of that share to deliver the share when an investor approaches to buy the shares in question. In this connection, the trader typically performs prop trading, i.e. trading on its own account. For obvious reading, the trading function is a very stressful job as significant amounts may be both gained and lost in seconds.

Traders rely on a continuous update on individual shares from equity analysts to buy or sell quickly and advantageously, as well as react commercially correctly when price-affecting news emerges. In addition to receiving formal research reports and e-mails, investment banks also have an internal live chat to receive instant news from brokers, traders and equity analysts. Sales and trading also have a constant live phone line across its offices to swiftly react to clients' incoming trading orders.

Economists and Strategists

The equity analyst is in constant dialogue with the investment bank's economists and equity strategists. The purpose is to exchange knowledge and views on macroeconomic developments. It is also to consider the overall investment strategies recommended to the investment bank's clients.

These recommendations include how much (proportionally) of the investors' wealth should be invested in equities and in which sectors (industries) it should be invested. As a rule of thumb, pharmaceutical and food companies ("non-cyclical companies") are more resistant to major stock market fluctuations in times of economic downturn, while companies that produce durable goods and materials ("cyclical companies") usually experience the largest relative stock price fluctuations during economic upswings. In conclusion, every sector has its own behavioural pattern, and generally, the investment industry often differentiates between cyclical and defensive sectors

The Equity Analyst's Working Day and Private Agenda

In our opinion, it is important for the collaboration between the company and the equity analyst that the company has insight into what the analyst's working day looks like. But it is also important to know what characterises the equity analyst's private agenda. To achieve such an understanding, one must initially have insight into the general structure of the equity analyst's salary. It should be emphasised, however, that the below description may differ depending on country, and that individual investment banks as well as local legislation in individual countries may dictate rules, restrictions and limitations on the structure of the salary packages of equity analysts (Fig. 10.3).

An equity analyst's total salary package often consists partly of a fixed salary, including pension, and partly of an individual discretionary bonus typically communicated in February or March. This bonus is partly based on the result that the investment bank has achieved in the previous financial year and partly on the individual performance of the equity analyst.

The bonus for the most successful equity analysts—especially in good times—can significantly exceed the size of the fixed annual salary and, in quite a few cases, amount to two or three times the fixed annual salary (in London and New York, higher multiples are seen), although this be caped due to local legislation. In other words, bonuses make up a very significant part of the analyst's total remuneration. There is a trend, however, that the bonus element of an equity analyst's total salary package is trending downwards as a consequence of regulative and political pressure, a general increased pressure

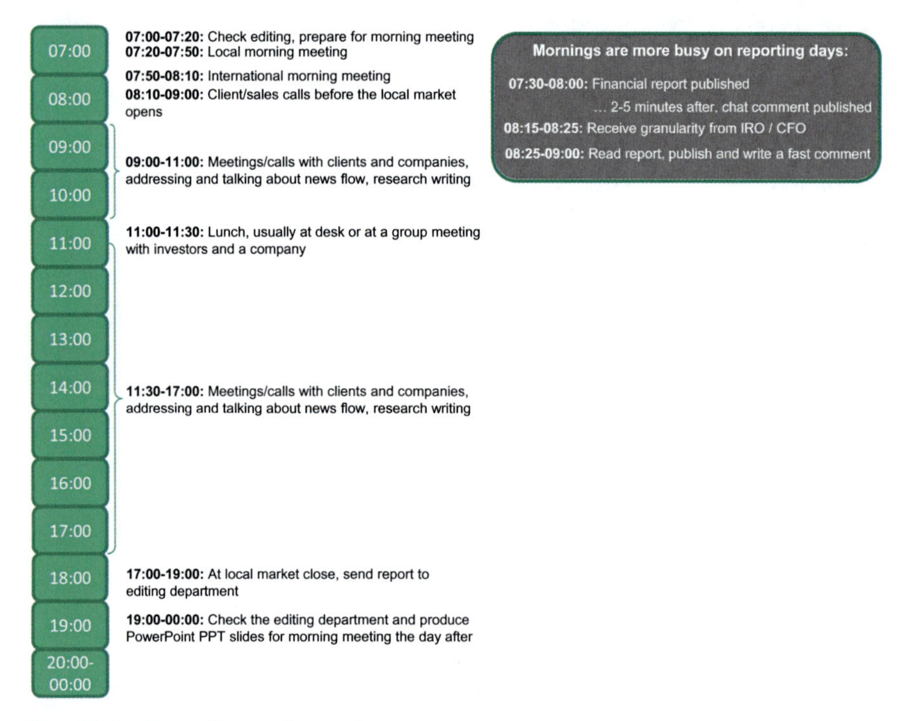

Fig. 10.3 Day of an equity analyst

on investment banks' commission-based earnings and a general trend towards rewarding the team and not the individual (at least to the same extent), which is also seen in other areas of society.

The bonus is awarded partly based on the result that the investment bank has achieved in the previous financial year and partly based on the estimated market value of the analyst in question, including the equity analyst's individual performance. To a very large extent, the equity analyst's market value is based on the equity analyst's ratings, corresponding to ratings the equity analyst has obtained from the institutional investors in connection with broker reviews. The overall rating of a broker consists of many detailed assessments in the following areas, among others:

- **Research,** including maintenance research, innovative research, sector research, macroeconomics, asset allocation, telephone service, scope and quality of arranged events, quality of recommendations, the scope of the investment bank's coverage of SMEs, quality of website and databases, etc.
- **Sales,** including overall customer service, morning update of customers, quality, etc.
- **Trading,** including sharp (advantageous) prices, flow (the volume that takes place through the investment bank, and which thereby gives the investment bank a sense of market development), market making (price position in shares), market insight, etc.
- **Back-office,** including low frequency of error notaries, same day handling (trading notaries, i.e. documentation, are sent out on the same day as the share trading is made), etc.

These ratings are weighted to an overall rating of the investment bank. Based on the overall rating, the institutional investor in question allocates a percentage of the total trading of shares to the broker—equity research weights relatively high (we estimate around 50%) in the overall rating. Therefore, the rating of the individual equity analyst has a high influence on the investment bank's overall rating and thereby the total commission that the broker is allocated each year.

The equity analyst is assessed for their ability and their value-added. This is a combination of the equity analyst's contribution regarding both their sector and the individual shares that the analyst cover. Against this background, the institutional investor indicates whether the equity analyst is number one, two or three, etc., competing equity analysts in the same sector.

The major institutional investors conduct these evaluations, or broker reviews, as they are also called, between one and four times per year. Therefore, equity analysts are always aware of spending their time on something that scores points with investors. These actions include:

- Write quality research reports.
- Bring the investor in contact with the company, i.e. facilitate relevant meetings between the company's management and the investor, including hosting the company's roadshows.
- Arrange professional events with, e.g. sector experts or peer companies.
- Be first with the news and especially a qualitative assessment of the possible consequences for the company in question.

In summary, an equity analyst score points and thus obtains ratings from an investor. It is based on their ability to enable the investor to make better investment decisions, which in turn should result in a better return on the investor's securities portfolio.

Furthermore, it must be noted that there is intense daily competition between equity analysts from the competing brokers, for the benefit of both the company and the investors. Therefore, equity analysts must very carefully prioritise their time in daily life. It is therefore clear that the actions which the company has a direct influence on, and which have a direct positive influence on the equity analyst's bonus include securing the following for the investment bank:

- Roadshows and events via the analyst's investment bank (leads to increased points for investors and thereby an increased analyst rating).
- Share buyback programmes through the equity analyst's investment bank (leads to increased commission income for the investment bank).
- Assist in securing advisory business for the investment bank's corporate finance department related to IPOs, secondary offerings (an already listed company issuing new shares) and M&A. As the equity analyst is independent, they are not remunerated directly for this point.

The Daily Relationships—Telephone Contact and Meetings

The company will typically, and hopefully, have good and frequent contact with the equity analysts who cover the company. This positively indicates an interest in the company from both the analyst and the investors. However, as we have mentioned before, equity analysts are reluctant to spend time on a company unless there is money associated with it, i.e. investors having a potential interest in trading shares in that company. Therefore, the company's contact with the analyst will typically be telephone contact, where the analyst regularly calls the IR function to get an in-depth comment on, for example:

- The company has issued a stock exchange announcement.
- A stock exchange announcement or press release has been issued by a peer company.
- Other matters of relevance to the company, i.e. the development of the company's share price.

Rarely, e.g. in connection with the preparation of major research report on the company, the equity analyst may wish to meet the company, either the IRO or management, for a more thorough update regarding the company's situation and conditions.

The dialogue between the company and the equity analyst must not naturally violate the relevant stock exchange rules regarding the disclosure obligations of listed companies.

This balance can be difficult to find, especially for newly listed companies on the stock exchange. The result is, sensibly enough, that these companies operate with a certain wide margin in respect of the stock exchange's rules on disclosure obligations of listed companies. One may naturally say "better safe than sorry" but for the equity analyst, this can, of course, be a bit cumbersome because the company does not seem very cooperative. But if the company is to avoid an unintentional offence, it must naturally be careful and get used to its new role as a listed company. This initial period of getting used to be listed is typically well understood by equity analysts.

There is no doubt that the equity analyst, like a skilled business journalist, goes after a good story and seeks to uncover matters that no one else has touched before.

The company must never expect that a piece of given privileged information will be kept confidential between the company and the analyst. It is simply illegal in the case of price-sensitive information. And it can be fateful for all parties involved if this information comes out in the market. This applies to the company, the equity analysts and any investor who may trade the company's shares with or without knowledge based on the information in question. If ever an unintended mistake should happen, it is recommended to quickly involve both relevant legal counsel and compliance to seek advice, prior to communicating any further to third parties. Then, matters will be sorted out, and appropriate action may be taken.

To ensure some consistent routines, the IRO must clarify to the equity analyst that the company's IR function is the equity analyst's primary and, as a starting point, the only entrance to the company. If the equity analyst has a close relationship with the company's management, and management does not mind, it is in order for the equity analyst to contact management directly. It is just a question of the company's policies and wishes in this area. Separately, what the company must take great efforts in avoiding is that the equity analyst, as can also be the case with journalists, explores the company and contacts employees further out in the organisation to dissuade them for information. Unless the employees are specially trained to deal with IR or equity analysts, it is the rule rather than the exception that the equity analyst will succeed in getting the employee in question to disclose potential price-sensitive information. Hence, relevant agreements must be made with the equity analysts, and company must create and enforce relevant internal communications policies.

Similarly, as a rule, journalists should contact the company's communications department or communications manager. Also, here, relevant agreements and communications policies should be in place.

The company can benefit greatly from the equity analyst. This is related to, e.g. competitor intelligence and improving the company's IR function. Good relationships with equity analysts can be maintained by having a constructive one-on-one relationship and inviting the equity analysts who follow the company to an annual lunch for an informal meeting. Such initiatives can be distributed throughout the year outside the company's blackout or frozen zone periods, which are the periods up to the company's quarterly result announcements.

In this connection, the management and the IRO may benefit from encouraging the equity analyst to prepare some thoughts and ideas regarding:

- How the company can improve its IR activities?
- How the equity analyst sees the competitive situation, including future trends in the industry?
- What initiatives would the equity analyst take if the equity analyst is "CEO for a day"?

Of course, the analyst does not have to decide anything, and the analyst is certainly not always right either. But the best equity analysts have considerable industry and competitor knowledge, and they closely follow the company. Furthermore, they are typically very passionate about their task and naturally want to impress the company's management with some creative thoughts.

The management of some companies will be grateful to hear what the equity analysts have to say about the company. Our opinion is that most companies can benefit positively from such brainstorming.

Corporate Access: Company Roadshows

All brokers and their equity analysts are typically queuing up to arrange the roadshows of major listed companies. This is mostly true for larger companies, where it is easy to attract investors; however, SMEs have found it more difficult to have roadshows arranged for them by the sell-side. Roadshows are allocated on a rotating basis to the various stockbroking companies for the largest listed companies. Usually, one would think that there will always be investors interested in attending a company presentation of its investment case. However, investors receive a significant number of invitations to investor presentations from equity analysts (or via the broker's corporate event team), who service the investors and who compete for broker review "points" and ratings. Investors must therefore prioritise their time, and sometimes the analyst must make a major effort to attract investors to an investor presentation.

This relationship underpins the main motivation for all IR activity. The company competes fiercely for the favour of investors to attract their attention. As a result, investors' money is invested in the shares that appear most attractive, and it is the shares that, due to increased demand, reduce the risk premium.

Some value investors like to look for undervalued shares that live their life in secret. But as a starting point, one must expect to be overlooked if one lives in hiding. It is especially companies in strong growth, which are expected to be dependent on financing via the financial markets, that need a clear profile in the market.

We recommend that the company is critical when choosing a broker for its domestic and international roadshows. The following considerations are recommended in connection with equity analysts, roadshows and investor meetings:

- Do not always favour one or two brokers or equity analysts. It is not always the perceived best, who is actually the best, and it is demotivating for all the other equity analysts if they can see that no matter how much effort they put in, it is the same analyst who always obtains the company's roadshows.
- Choose a broker based on who is creative in arranging investor meetings with high-quality investors and who the company has not met with before.
- Choose a broker who represents a particular strength in each geographical area where the roadshow will take place and ask the broker about their geographical strengths and weaknesses. If the broker commits to a certain geographical area, they are also forced to perform well—otherwise, they will later be punished accordingly in terms of either fewer or no allocated roadshows.
- Choose a broker according to who in the previous period, e.g. financial year, has made an extra effort in the company's marketing or who has prepared one or more major analysis reports on the company.
- Do not blindly opt out of the equity analysts who have a negative recommendation on the company's shares. Incidentally, this typically makes the analyst even more negative on the company. Equity analysts' negative views may well be valid, and no one knows the future after all. By going to a roadshow with one of the equity analysts who has a more negative view of the future of the company's shares, the company gets a good opportunity to challenge the analyst's views constructively.
- Do not expose or ridicule an equity analyst in the eyes of investors. It does not belong anywhere, and investors do not like that a creative analyst, who has reached an unpopular conclusion regarding a listed company, is punished for certain thoughts which may even later turn out to be a correct conclusion.
- When travelling around for several days with an analyst at a roadshow abroad, the company's management spends a lot of time with that analyst. Be careful not to open the bag of confidential information gradually. The equity analyst is often good and well trained at "opening up" people and, as a rule, always has one thing on the mind, which is completely legitimate for an equity analyst—to come home from the trip with satisfied

investors and a little extra information that the equity analyst can share with both investors, and internally with sales and trading.

- Never allocate a roadshow in the same city or geographical area, during the same period, to more than one broker. If two different brokers may call the same investor to set up a meeting with a company on the same day, or days, the brokers will not become amused, to put it mildly.

The company typically pays for its own airline tickets and hotel stays. In terms of practice, it is customary for the broker to pay the fee for holding roadshows, including premises for the presentation, travelling between meetings and catering.

The IRO must map the quality of each roadshow. It is therefore useful to develop an internal evaluation tool (see example below) based on areas of evaluation and a score system from 1 to 5 (where five is best). The broker can thus achieve a maximum of 5.00 points. This is a useful tool in order to better allocate roadshows in the future as historic performance is a useful deciding factor. It is important to include the quality and continuity of equity research, as this contributes to the overall perception of the roadshow.

Investor roadshows – evaluation of broker performance

No.	*Areas of evaluation (suggested)*	*Suggested weight (%)*	*Points (1-5; where 5 is best)*	*Points x Weight (Two decimals)*
1	Quality of investors met	25%		
2	Relevant number of investors met	10%		
3	Client profiles (written investor descriptions)	5%		
4	Pre-roadshow communication with broker	5%		
5	Logistics (planning of the day)	10%		
6	Interaction with the broker during the roadshow	5%		
7	Quality of feedback reports regarding investor meetings	15%		
8	Continuity of company research coverage	10%		
9	Quality of company research coverage	15%		
	Total	**100%**		

Reference and Source

1. Selected illustrations in Chapters 2, 4, and 5 are inspired by: Lykkesfeldt, P. (2006). Investor relations: A book about stock market participants. *Børsen*.

CHAPTER 11

Corporate Finance Advisers

Introduction to Corporate Finance

For the past decades, in particular starting in the early 1990s following a series of insider scandals around the world, the rules regarding contact and exchange of information between the corporate finance department and other departments, including equity analysts, have been significantly tightened up via a significant expansion of the compliance function. Certain rules existed prior to the early 1990s, but they were not enforced in the same manner as it is the case today. Compliance rules are internal rules within of investment banks that seek to ensure correct business and ethical conduct within the existing legal framework. These austerity measures, an international phenomenon, have gradually developed and increased during recent years. As a result, equity analysts' advisory behaviour is regulated by conflicts of interest, including in connection with the equity analysts' possible collaboration with the investment bank's corporate finance department. As corporate finance is working with companies on, e.g. raising capital or strategic advisory, the investment bank needs clear Chinese walls to block the exchange of information from the corporate finance department and other departments and functions of the investment bank.

The situations where an equity analyst has contact with the corporate finance department typically include IPOs and in connection with listed companies' issue of new share capital (new issues or secondary offerings) or M&A transactions. In these situations, there are strict established rules with compliance oversight for how the equity analyst should act towards investors, and if and when the equity analyst should "be brought over the wall".

Unless the equity analyst is "brought over the wall", and then per definition becomes an insider, communication between the equity analyst and the

P. Lykkesfeldt and L. L. Kjaergaard, *Investor Relations and ESG Reporting in a Regulatory Perspective*, https://doi.org/10.1007/978-3-031-05800-4_11

corporate finance department is characterised by one-way communication only when corporate finance employees occasionally seek general knowledge about a given industry without revealing why they are interested in this knowledge. On the other hand, naturally, the analyst cannot seek information from the corporate finance department.

Corporate finance is engaged by, e.g.:

- A company approaching a corporate finance department directly.
- A company issues a request-for-proposal ("RFP") to receive proposals from different investment banks regarding a contemplated transaction.
- A company is approached directly by a corporate finance department, who considers an idea and wish to pitch the idea for the company.
- A financial advisor (e.g. the company's lawyers) approaches a corporate finance department on behalf of a company as their client.
- An internal analyst notifies, subject to local regulations, corporate finance of an idea they may have discovered—and where the idea may later materialise. However, the representatives of corporate finance and equity research must always be aware of relevant compliance in the given country and consult the relevant compliance function if in doubt.

Relationship with Corporate Finance Advisers

Often, the company's CFO has primary contact with a corporate finance department. In particular, larger listed companies regularly receive inquiries from corporate finance advisers from domestic and foreign investment banks about ideas and proposals for transactions. Other companies automatically choose their regular investment bank to solve a specific corporate finance task, although the trend is definitely to have several corporate finance relationships. Listed SMEs usually have practical experience with corporate finance advisers from their IPO or a subsequent increase in their share capital. They may also have completed an M&A transaction. Several listed companies receive inquiries from corporate finance advisers from time to time. The purpose is often to keep in touch in order to develop a preferred relationship with a view to obtain a future mandate.

There is no doubt that collaborating with corporate finance advisors can be lucrative and profitable for a company. The company can often pick up a large amount of advice at no charge if it plays its cards correctly. Conversely, corporate finance advisers are also experts in maximising its own profits. If the company is under pressure, they often have the upper hand in negotiating fees and fee structure with the company. They have this, partly because there is a large degree of opacity in the fee structures and partly because the company often turns to corporate finance advisers in the eleventh hour, where the company only has a few options about the choice of corporate finance advisor. Finally, often, the company is also inexperienced in negotiating fees with corporate finance advisers and is not fully aware of its alternative options.

Companies' experiences with corporate finance advisors vary widely, ranging from the situations where the corporate finance advisor is essential to creating value for a given transaction to the cases where the corporate finance advisor's contributions are less visible. But that does not have to be the case. For many companies, it is a matter of lack of insight and preparation, too few demands on the corporate finance adviser up-front, and a reluctance to set the agenda in connection with a planned or a current transaction. The company should remedy this. One should not be mistaken, used in the correct manner, a corporate finance may contribute significant value to a company.

This chapter aims to illustrate how the corporate finance advisor works, where the corporate finance advisor seriously creates value for the company, how the company best works with a corporate finance adviser, and to seek to point out how the company avoids paying more than necessary for good corporate finance advice (Fig. 11.1).

Fig. 11.1 Stakeholders of corporate finance

Scope of Corporate Finance Advisers

In addition to the above parties, several specialists may also be involved on an ad hoc basis, e.g. in connection with the due diligence procedure, depending on the nature of the transaction and the industry concerned. However, as a rule, the corporate finance advisor plays the important role of project coordinator.

The corporate finance advisor may seek to minimise its own efforts, however not on the expense of the quality perceived by the company. In addition to its reputation, capacity, resources and time are the things that the corporate finance advisor guards the most. The corporate finance adviser naturally also seeks to ensure that the transaction becomes a reality. As a rule, the corporate finance advisor works according to the no cure, no pay principle, i.e. no transaction, no fee.

In the following, we look at the areas in which the corporate finance adviser is particularly justified and makes a significant value-creating contribution (Fig. 11.2).

Project Coordination

The corporate finance advisor typically handles the project coordinator role in connection with an IPO or a M&A transaction. In corporate finance-related transactions, there is often a large element of tactics and timing. Furthermore, there are often many parties involved. For example, in the case of an M&A transaction, there are advisers and experts on both the buyer's and seller's sides. Therefore, one of the important tasks of the corporate finance consultant is, as a rule, to coordinate the given project to ensure the greatest value creation for the client, i.e. the company.

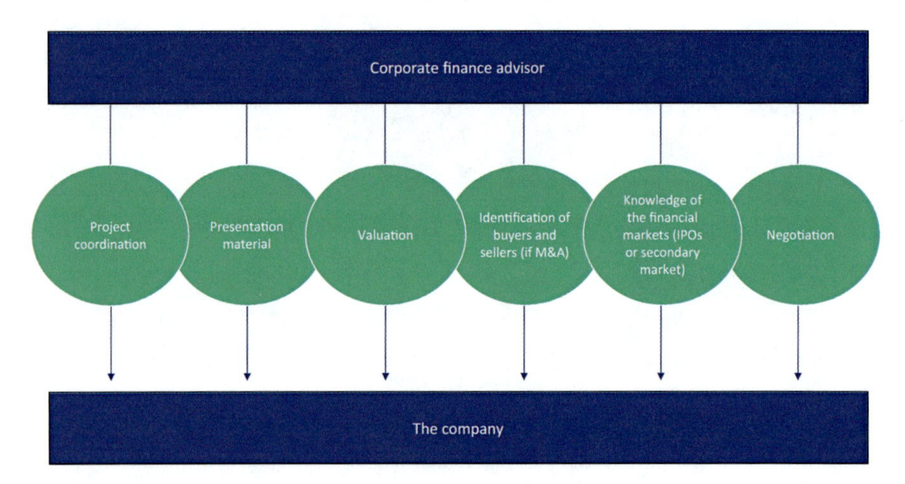

Fig. 11.2 Unique selling proposition of corporate finance

Presentation Materials

As is often the case with advisors, corporate finance advisors are very adept at preparing illustrative and compelling presentations. Through these, among other things, the corporate finance adviser initially seeks to win their pitch in connection with a possible beauty contest, where several corporate finance advisers present themselves intending to win an advisory assignment. Most often in competition with other corporate finance advisers.

The corporate finance adviser's ability to present and structure is also expressed in connection with the preparation of a so-called information memorandum, which is the document that describes a company or business unit (BU) that may be sought divested. Finally, the so-called prospectus is prepared in connection with an IPO or a share capital increase (new issue or secondary offering). It is the company's ultimate responsibility to ensure the correctness in all presentation and prospectus materials. The corporate advisers and other consultant may provide advice; however, the company is responsible.

Valuation

The Corporate Finance Advisor's Expertise in Valuation Plays a Critical Role

It is seldom possible to point to a company's or a BU's precise value. This is based on a series of assumptions. The role of the corporate finance adviser is to present their views on valuation in the specific case to the company's management, and later a possible counterparty of the transaction, as professionally and convincing as possible and then back up their views with their good name and reputation.

In price negotiations, whether it is an IPO, an M&A transaction or a third-party takeover bid for the company's shares, it is essential to have as good a corporate finance advisor as possible to achieve the best price on behalf of the company for the benefit of its shareholders.

The valuation techniques, i.e. DCF and the comparison of financial markets-related key figures for comparable companies, are basically in line with those used by equity analysts. This is in particular the case in connection with IPOs. However, although corporate finance advisers use the same overall valuation techniques as equity analysts, there is often a different approach to choosing comparable companies.

While equity analysts typically focus on listed peer companies, aiming to value shares in one of the companies the analyst follows. However, in their work with M&A transactions, the corporate finance adviser seeks to value both listed companies, but often to an even greater extent the terms and conditions, especially relevant valuation multiples, on which both listed and non-listed companies were actually acquired or merged.

Identification of Buyer/Seller (M&A) and Knowledge of the Institutional Investors

Another important role of the corporate finance advisor in connection with M&A transactions is identifying potential buyers and sellers. When seeking to divest a company or BU, it is essential to ensure that as many potential buyers as possible are identified from the outset. This is to assess which potential buyers should initially be contacted by the corporate finance advisor.

In other contexts, the corporate finance adviser from an investment bank may search to identify companies or BU that could be potential acquisition candidates for the corporate finance adviser's client.

Knowledge of the Financial Markets (IPOs and M&A)

For a listed company, the financial markets will always have an opinion on a corporate finance-related transaction. Therefore, the corporate finance advisor must assess how investors and equity analysts will react to a given transaction. As previously mentioned, the company should not be dictated to by the financial markets. Still nevertheless, the company needs to know how the financial markets will react to a given, intended transaction—before its completion.

Another factor that the corporate finance advisor, for natural reasons, is concerned with and has insight into is the significance of a given timing for a proposed transaction. The financial markets' expected response to an intended IPO, share capital increase or M&A transaction is thus a critical element in the corporate finance context.

In connection with IPOs, the strengths of the corporate finance advisor also include a detailed knowledge of stock exchange legislation, the financial market dynamics, as well as the ECM team's good knowledge of relevant institutional investors.

Unfortunately, however, seldomly, it is seen that the judgement of corporate finance advisers regarding timing, valuation or business rationale may be weakened by the desire to win the mandate for a given advisory task, even though this should not be the case with professional quality corporate finance advisors, who focus on adding value and building a long-term relationship with the company.

Negotiation

The last and critical phase in which the corporate finance adviser is involved and sometimes primarily responsible is the negotiation phase.

The corporate finance advisor is typically a trained negotiator, and in that phase, this can be extremely valuable to their client. In addition, a skilled corporate finance advisor can easily earn a multiple of their fee back home to the client—alone in this phase.

Of course, the company must also be aware that the corporate finance adviser's experience and abilities in negotiations are why the company often feels that corporate finance advisory agreements are entered into for a relatively high fee. In this context, the corporate finance adviser may refer to the fact that the advisory fee must be seen in connection with the value the corporate finance adviser adds to the transaction.

The Corporate Finance advisor's Fees

The corporate finance adviser is generally remunerated with a transaction-based fee, i.e. no cure, no pay fee related to a percentage of the transaction value. However, there may also be an additional retainer fee, e.g. a fixed monthly fee. This part is very limited in comparison with the total potential fee if the transaction is completed. Finally, there may also be a discretionary, or non-discretionary, success fee.

Therefore, it should be obvious that the corporate finance adviser is very transaction driven. However, the corporate finance adviser is, of course, obliged to advise in the interest of their client. As is the case with the equity analyst, the corporate finance adviser's salary structure is characterised by a fixed salary and a bonus that relates to the corporate finance department's or the total investment bank's financial result for the previous financial year, as well as the individual performance. Bonuses are typically paid out in February or March. The level of bonus that successful senior corporate finance advisers may receive is, subject to a number of factors, often at a higher level, relative to the fixed salary, than the equity analysts, provided the same experience level. This relationship helps put the corporate finance adviser's focus on implementing a proposed transaction in perspective.

The Daily Relationships

Like so many other areas of the business world, the award of a mandate in connection with a corporate finance task is often based on good relations between the company and the corporate finance advisor.

Therefore, the corporate finance advisor seeks to stay close in touch with the company for good reasons. This can take place either through informal meetings (this is often the case when the relationship is already reasonably close) or by the corporate finance adviser contacting the company to meet about an idea, a theme or something else that the company must be presumed to find worth meeting about.

When choosing a corporate finance advisor, it often takes place in the following circumstances:

- Before an intended IPO, divestment of one or more BUs, acquisition or merger, the board of directors and senior management will often invite two to four investment banks to pitch on the proposed transaction. Based on this pitch, a sales presentation and separate discussions with the relevant investment banks, the company will select one advisor to solve the task. This is often also the case where the company has a regular corporate finance advisor, which solves most of the company's corporate finance-related transactions.

- In other cases, the corporate finance advisor brings an M&A idea to the company. If this is new and innovative for the company, and the company chooses to pursue the investment bank's proposal, it is customary for the company to choose the corporate finance adviser who presented the idea first. However, fees must first be negotiated.
- In an expected or actual bid for the company's shares by a third party, there is a need to act very quickly. In these cases, corporate finance advisor is chosen very quickly. This choice is usually based on existing relationships and a sense of which corporate finance advisor can solve the task best. In such situations, there is often no time for a formalised beauty contest (it is in such situations that the company benefits greatly from having already established a contingency, including an up-to-date takeover response manual).

It is generally recommended to meet from time to time with corporate finance advisers from investment banks, who have made a good impression, to establish some relationships that the company can benefit from, the day the need for a corporate finance advice suddenly arises—and decisions have to be made very quickly.

Corporate finance advisers, like equity analysts, can often provide a good update on what is happening in the industry as they are typically quite insightful, intelligent and have good access to many sources of information. Even if the company is naturally very well informed, the financial advisers often have some interesting and significant nuances, which can be valuable knowledge in connection with the company's competitor intelligence. In addition, the corporate finance adviser may bring a concrete M&A idea that the investment bank is aware of or has a reasonable suspicion that could become relevant.

Thus, it is often possible to obtain valuable input to supplement the company's knowledge base on the expected future industry development and competitive situation.

Selection of Corporate Finance Advisors and Control of the Transaction Process

The number of transactions in which investment banks have advised companies has risen sharply for several decades. This is especially true in the M&A area, both because of the companies' expansion or restructuring, e.g. divestment of non-core areas or loss-making activities, and the increased activity of private equity funds.

We believe that in some cases, the selection of investment banks and corporate finance advisers takes place with an element of randomness and lack of systematics. Furthermore, it is our view that considering the importance for the company of a typical transaction, and the often-significant fees that corporate finance advisers receive for their services, significant value may be created for the company if this process is carried out by the company in a more structured and systematised way. This applies to the phases:

- Selection of investment bank.
- Negotiation of terms.
- Project follow-up.

Be Out Well in Advance and in Good Time

If this is practically possible, it is advantageous for the company to be out in good time when choosing a corporate finance advisor. The more time the company has, the better the negotiating situation the company is in.

Since there is time to conduct simultaneous discussions with more than one corporate finance adviser, the proposed transaction can be better considered before significant decisions are made.

Clarification of Assignments

We recommend always entering a dialogue with at least two corporate finance advisers before a final choice. This is true even when the company is in a special situation. However, this does not apply if the company feels confident in its choice of corporate finance advisor. Two rather than one corporate finance adviser are recommended to solve the question from different perspectives at an early stage.

In the subsequent pitch (the sales-oriented presentation by the corporate finance adviser to the company), the corporate finance adviser will seek to set the agenda for what the company should focus on and what should take the most space in the presentation. It is recommended that the company sets the agenda by preparing a specification of requirements, and issues to be addressed, for the corporate finance adviser to make the various pitches. The company must compare them and ensure that the corporate finance adviser addresses the areas where the company wants the position, views and recommendations of the corporate finance adviser. For example, this requirements specification could include requirements for:

- What should the pitch contain?
- What issues should be addressed?
- What structure should the pitch have?
- What is the corporate finance adviser's specific proposal for fee structure, including varied and fixed fee types, as well as a possible discretionary success-related fee?

Some of the areas that, depending on the type of transaction, should be given extra weight include:

- The chemistry between the company and the corporate finance advisor.
- Assurance that the corporate finance team presented is also the company's team throughout the transaction.
- The level of experience of the corporate finance team members with whom the company has the primary day-to-day contact throughout the transaction.
- The investment bank's experience and strength within the relevant type of corporate finance transactions in the industry in question, i.e. track record within comparable transactions.
- The investment bank's strength (rating) within equity research in the industry in question.
- The corporate finance adviser's competence in distributing shares in the industry in question in the event of an ECM-related transaction, i.e. an IPO, a secondary offering, or a capital raise for an unlisted private company.

Negotiation of Terms

Once a corporate finance advisor has completed a pitch for a given transaction, the corporate finance advisor generally has:

- Been through a very extensive research process.
- Been through the process mentally.
- Completed, albeit preliminary, significant valuation work.
- Prepared a larger presentation material.

It is worth noting that significant sunk costs, i.e. costs that are immediately depreciated, have therefore already taken place for the corporate finance adviser at the time corporate finance adviser enters the meeting with the board of directors and/or senior management, to deliver the pitch for the advisory task.

Once the corporate finance adviser has completed its pitch and a subsequent discussion of the intended transaction has taken place, the company must make its choice. The corporate finance adviser's fee is an element of this decision-making process for the company, but most definitely not everything.

The company should be fully clear from the outset at the time of discussion and negotiation of transaction fees:

- There is typically different profitability for the corporate finance advisor depending on the type of transaction. For example, an IPO is more labour intensive than a more straightforward M&A transaction.

- A corporate finance advisor has a willingness to work very hard and long and has almost per definition always available capacity as the always use all of the hours of the day (and night).
- In the case of some transactions, there is considerable prestige associated with the corporate finance adviser. This is because these transactions can be counted in various rating tables, which can help the adviser to win future assignments. Further, completed transactions are widely used by the corporate finance adviser in its marketing for new assignments.

Thus, there is often far greater scope for the company in negotiating terms with the corporate finance adviser than the company is immediately aware of.

The company must choose the corporate finance adviser with whom the company feels most comfortable also at the personal level, but at the same time, it is often possible to hold back a little with the shareholders' money. However, the company should be aware that the corporate finance advisers are experienced negotiators. Hence, there are naturally limits as to how little the company can negotiate the fee down to. Also, the company for very good reasons has a wish for the corporate finance advisor to stay committed and enthusiastic throughout the transaction.

Evaluation and Selection of the Corporate Finance Advisor

The company should include an evolution tool to be as simple as possible to use in connection with:

- The mutual comparison of the various corporate finance advisers.
- The evaluation of the individual corporate finance advisor.
- The final choice of corporate finance advisor.

In connection with the pitch of the various corporate finance advisers, it is essential, possibly through separate questions and conversations, to obtain a feeling of all the individual members of the team—not just one of the two senior people who speak in connection with the pitch. It is often the team's younger members that the company deals with the most during the transaction itself.

Prior to the final selection of corporate finance adviser, the company may assess the corporate finance adviser in a quantifiable manner in which ten sub-areas are evaluated with a score of 1–5 (where five is best). The score is weighted according to a relative materiality criterion. A corporate finance advisor can thus achieve a maximum of 5.00 points, where irrelevant areas are skipped and not included. The below table and weightings are naturally merely for inspiration and may be modified by companies. Finally, the conclusions of the evaluation overview are naturally not conclusive but should merely be seen as a helping hand to companies in their selection process.

Evaluation of a corporate finance advisor

No.	*Areas of evaluation*	*Suggested weight (%)*	*Points (1–5; where 5 is best)*	*Points × Weight (two decimals)*
1	How well did the corporate finance advisor follow the specified requirements in connection with its pitch?	10		
2	To what extent did the corporate finance advisor include irrelevant or misleading matters in its pitch?	5		
3	The investment bank's experience and strength within the relevant type of corporate finance transaction in the company's industry?	10		
4	The investment bank's strength and rating within equity research in the industry in question?	10		
5	The corporate finance adviser's competence and strength in the distribution of shares in companies in the industry in question (if relevant for the type of transaction)?	10		
6	The corporate finance advisor's understanding and innovative thinking in connection with the proposed transaction?	10		
7	Assurance that the presented corporate finance team is maintained throughout the transaction?	5		
8	Relevance and reasonableness of the proposed fee structure?	5		
9	The chemistry between the company and the corporate finance advisor, including the actual transaction team that is to represent the corporate finance advisor throughout the transaction?	20		
10	Overall professional impression of the competence of the investment bank and the corporate finance team?	15		
	Total	100		

The primary purpose of the above evaluation tool is to ensure that the company systematically gets around the corners in connection with the assessment of the individual corporate finance advisers and can document to itself why a given final choice is made.

Furthermore, we recommend asking the corporate finance adviser about previously solved tasks within the same area and contacting other companies that have had a transaction executed by the corporate finance adviser in question. The corporate finance adviser often publishes which transactions they have advised in connection with.

Management of the Transaction Process

In connection with the management of the daily transaction process, it is recommended to let the corporate finance advisor do as much of the work as possible. This is because, on the one hand, that most often work on a non-hourly fee but based on a fixed be, and on the other hand, they have extensive experience in managing the process effectively, and are typically very skilful.

It is recommended to show a healthy critical sense, as one should also do, towards the assessments and recommendations of corporate finance advisers, as they are, after all, typically quite transaction oriented due to the typical no cure-no pay fee structure.

If there is a need for more than one corporate finance advisor, e.g. in connection with the desire for a greater geographical spread of the share capital in an IPO context, a merger or secondary offering in connection with a large M&A transaction, a choice of a lead coordinator is generally recommended. A lead coordinator, the corporate finance advisor or investment bank is responsible for the overall coordination of the transaction, including the other corporate finance advisor(s). The alternative is an advisory structure, which often requires greater involvement on the part of the company, among other things, to clarify potential disagreements among the members of the consortium.

CHAPTER 12

Considering the Role of Non-financial Markets Stakeholders

External Stakeholders

In addition to considering the stakeholders of the financial markets, a listed company should also consider how they communicate and position themselves towards non-financial stakeholders. We have entered the "information age" where today information is alive. For obvious reasons, for any company, it is essential both to the company's stakeholders and to map the issues relevant versus all the company's stakeholders. On this basis, the company may develop an issues management contingency.

In addition to the more perhaps traditional stakeholders and issues, a company is also well-advised to consider historically overlooked matters, which today is top of the agenda. These include (not exhausted) the #MeToo campaign, social activism, employee rights, political correctness, racism, gender, inclusion, etc. Not taking these issues seriously may have severe repercussions by society and overall social sentiment towards a company. In addition to hurting social sentiment and increasing the risk premium of a company's shares, a company can also damage its standing towards suppliers and customers and thereby their profitability. It is, therefore, also in the interest of investors that a company has a comprehensive mapping of its non-financial markets' stakeholders.

In the below overview, we have mapped a series of the company's traditional stakeholder. However, the stakeholders included in the illustration, as well as the distribution between primary and secondary stakeholder may vary from company to company (Fig. 12.1).

As listed companies pursue transparency to conduct best-practice IR in addition to the general acceleration of digital communication and social media, non-financial external stakeholders have only been an increasing force. Like the

P. Lykkesfeldt and L. L. Kjaergaard, *Investor Relations and ESG Reporting in a Regulatory Perspective*, https://doi.org/10.1007/978-3-031-05800-4_12

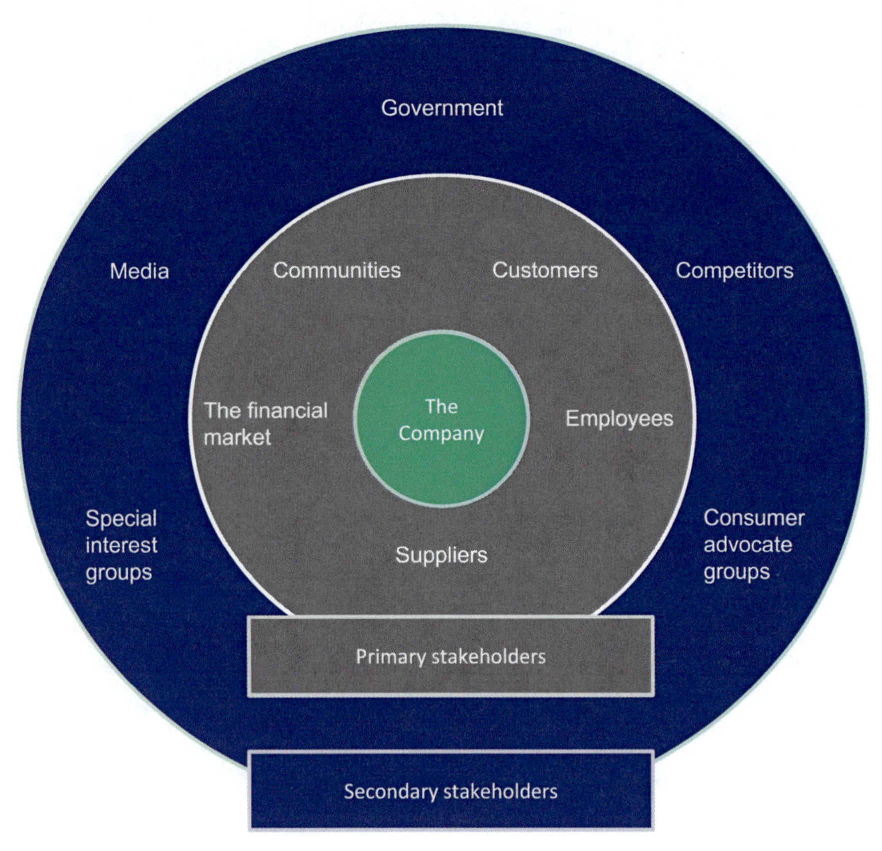

Fig. 12.1 Company's stakeholders

company's communication with retail investors, non-financial external stakeholders have today obtained a more robust and meaningful platform. In the past, it was common for companies to have "quiet" days with limited news flow and therefore limited trading in the shares.

Today, news flows of different nature every day impact the volume of shares traded. Therefore, the IR function needs to plan and have guidelines as well as an issue management contingency. Further, it is important to align internal stakeholders to be aware of the company's ESG guidelines and how any unethical business may have severe consequences for society and the company's overall objectives.

Company Competitors

As a listed company, it is expected and required to have a high degree of transparency. There are many excellent books on corporate strategy and competitive advantage. For example, Professor Michael S. Porter defines two ways an organisation can achieve a competitive advantage versus its competition: cost and differential advantages. However, the topic is incredibly vast and challenging with respect to information sharing. Risks to a company for not obeying competition law include fines, disqualification of senior management, reputational damage, prison and criminal penalties for individuals involved.

The OECD's general standpoint[1] is that information exchange between companies and their customers, suppliers and even competitors increases transparency in the market. However, despite enhancing the overall market efficiency, it can also present competition risks. Transparency can lead to collusive equilibria among competitors due to non-coordinated anticompetitive effects. Competition law is mainly concerned with strategic information. Examples include providing information on prices, volumes or capacity, artificially increasing transparency in the market, and allowing collusive behaviour as key focal parameters of commercial competition.[2]

Therefore, companies must disclose sensitive market information publicly and balance it with their ongoing business activities in a competitive environment.

Media Relations

Typically, the functions of IR and media relations/public relations in a company are separated. The reason is that the communication with the media differs materially with the communications with the financial markets.

Media relations involve working with the media to inform the public of a company's mission, policies and practices in a positive, consistent and credible manner. In this connection, any company information is typically widely spread in the media.

In the IR function, any information, naturally of a non-price-sensitive nature, is typically distributed much less widely, if at all. The information is also often of a much more in-depth and detailed nature.

This typically requires different experience and backgrounds of the company individuals exercising their role.

In our experience, when the IRO discusses themes and topics with equity analysts and institutional investments, there is an understanding of the implicit messaging. For example, it is not beneficial for long-only institutional investors to share the competitive situation of a company. Therefore, there is a high consideration of the granularity of information that a company shares with its financial stakeholders.

Sometimes IROs also talk to the media; however, a best-practice IRO must distinguish how to disclose information. In essence, any word communicated to the media can be a top story later in the day as it is in the interest of journalists to write articles that create attention. Therefore, it recommended that, due to the different nature of the external stakeholders, and due to the relevant work pressure, the roles are separated at the operational level, unless there are special circumstances in the company, or the experience and professional background of the individuals performing the various tasks. However, communication with the media is not a topic of this book. Still, it is recommended, however, that the IRO has some kind of involvement in the work with the media, e.g. when very detailed and insightful journalists need to be serviced, as it is always useful to get an insight of other stakeholders than one is used to deal with.

A summary of facts and best practices is illustrated on page 329

References and Sources

1. OECD (2010). Information exchange between competitors under competition law. *Policy Roundtables, DAF/Comp (2010)37.*
2. Lourenco, N. (2018). Information exchange between competitors from a competition law perspective—the problem of premature exchanges of sensitive information in the context of merger control. 2018. *Unio EU Law Journal*, Vol. 4(1), pp. 78–101.

PART III

Major Legislation Themes Related to the European Financial Markets

CHAPTER 13

How Is Legislation Implemented on the European Financial Markets?

It takes less time to do things right than to explain why you did it wrong.

—Henry Wadsworth Longfellow, Poet

The EU's Regulatory Wave

This chapter discusses the most relevant regulatory themes within the legal framework of MiFID II[1] (like FINRA[2] in the US), MiFIR[3] and MAR.[4] These regulatory themes are extensive and vast. Our intention is not to exhaust the reader with an in-depth detailed review of legislation. We leave this for the distinguished members of the legal and audit profession.

Rather, it is our ambition to create an overview of the most profound current and coming regulatory framework relevant for financial markets practitioners and their daily work, and to highlight some of their key attributes relevant for the market participants. This includes providing vital cornerstones and how professionals operate within the regulatory framework to both secure that relevant legislation is fully complied with, and at the same time conduct efficient IR for the benefit of the financial markets' stakeholders.

Further, our intention is also to shine a light on alternative solutions and approaches that have sprung from the introduction of the regulatory themes. We believe that the IRO should have a solid understanding of the regulatory framework of the investment banks and institutional investors, as they are the most essential financial stakeholders of the company.

Legal scholars and the financial markets participants have coined the term "regulatory wave" to introduce new legislation following the 2008–2010 financial crisis. There have in essence always been relevant laws in place in all

P. Lykkesfeldt and L. L. Kjaergaard, *Investor Relations and ESG Reporting in a Regulatory Perspective*, https://doi.org/10.1007/978-3-031-05800-4_13

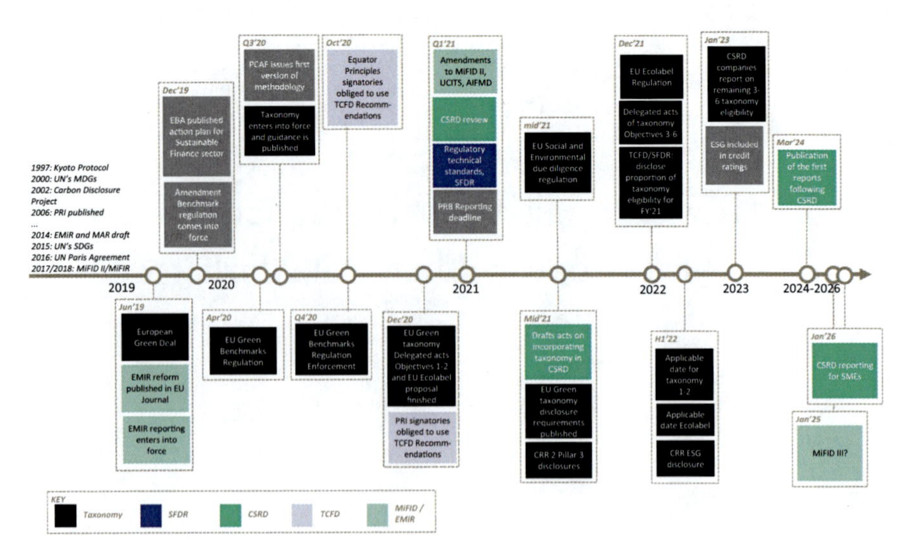

Fig. 13.1 Timeline of the EU's regulatory wave

well-developed financial markets. However, the amount, detail and comprehensiveness of the legislation accelerated following the financial crisis. Set out below is the legislation which today is central for all stakeholders operating on the financial markets as well as an overview of currently planned new legislation (Fig. 13.1).

The Context of Financial Legislation

Legislation is no longer an isolated discipline but has become a critical strategic dimension for financial intermediaries and companies competing in a globalised world marked by the increasing exchange of information. Therefore, legislation must be considered as a top management responsibility and priority. The applied legislation means increased transparency leading to the enforceability of compliance by regulators and investors/investors. In addition, there is a need in companies for the structuring of behaviour (processes, training and tools) and the creation of awareness and multi-disciplinary know-how throughout the entire organisation.

The regulatory process must in companies be holistically viewed and treated as an interdisciplinary effort linking legal with business to ensure compliance—right from interpretation to product/service structuring, and up to the delivery and behaviour at the investor front. The continued convergence of tax and regulatory worlds, coupled with conceptually different regimes applied in different parts of the world, multiplies complexity. It significantly raises the bar on all key levels of the value chain.

Timeline of Financial Legislation

All legislation of the financial markets aims to ensure and promote fair and transparent markets. We believe that availability of public discussion, and of relevant literature, regarding the financial legislation in the US is under development (see Chapter 40). We naturally encourage legislators in the US, and other jurisdictions, to embrace the initiatives taken by the EU. Therefore, in this book, we focus on legislation in the EU, which is largely aligned with similar laws in the US and other legal jurisdictions globally. In our experience, we believe this important theme disserves more emphasis and that it will only grow further in importance. We focus on the key attributes for practitioners, but when conducting business, senior managers and IROs are recommended to consult with internal and external legal counsel.

Introduction of Regulatory Frameworks

The law is heavily dictated by culture. Therefore, every country in Europe has different legal frameworks, laws and penalties. However, most of the countries are members of the EU. Therefore, the legal similarities are high. This is especially the case for laws on the financial markets. Fundamentally, EU legislators endeavour to create a legal framework that encompasses the aims of the EU treaties. Treaties, known as primary law, are the purpose and involvement of every action taken by the EU.

Primarily law is implemented by proposing and implementing several types of legal acts (legislation) known as secondary law. Some of the legislation is binding, some are not, some apply to all EU member states, and some do not—the main types are:

- **Regulations:** A binding legislative act for all EU member states and is automatically implemented into national law.
- **Directives:** Common legislative goals that all EU member states must achieve by implementing necessary acts into local legislation. Member states can simply implement the directives into their local law. However, directives allow national flexibility. Furthermore, the local implementation must be communicated to the European commission (generally within two years) or infringe proceedings.
- **Decisions:** A legal decision on certain institutions (specific countries, organisations or companies) that they must comply with.
- **Delegated acts:** A legal act that allows the EU Commission to amend or supplement non-essential EU legislation, for example, detailed measures.
- **Implementing/technical acts:** Pieces of practical legislation, supervised by committees, set conditions that ensure uniform implementation of the EU's legislation.
- **Recommendations:** A non-binding legal act on areas the EU believes should be considered, improved or reviewed. This allows making the

EU's considerations known and allowing the concerning institutions to navigate.

- **Opinions:** Like a recommendation, an EU institution (e.g. Commission, Council, Parliament or a Committee) can, in a non-binding fashion, provide an opinion which legislators and external institutions can consider and implement.
- **Q&A:** Like opinions, following the implementation of major legislation, institutions may submit public inquiries/questions about the practical implementation of the legislative acts. EU institutions (e.g. European Securities and Market Authority, ESMA) will respond to the questions with tangible considerations of implementing the legislation.

The EU's considerations are principal for most of the legislation in Europe. In addition, the considerations are widely aligned with EU/EEA member states and "third-country states", i.e. non-member countries, including the US and the United Kingdom. Therefore, our main suggestion for the IRO is to be familiar with the EU and US institutions' financial framework and local laws where the company is incorporated. However, most frameworks are highly aligned, and it is the responsibility of legal counsel to advise on deviations for the practitioners.

For the financial markets, particularly the scope of IR, the three most relevant legislations currently in the EU are MiFID II/MIFIR and MAR, which we intend to explore in more detail. First, however, we highly recommend that the IRO is be updated on all major legislation concerning the financial industry and it is important to note that these regulatory frameworks are constantly being developed.

MiFID II/MiFIR and MiFIR

MiFID was introduced in 2007 to protect investors, increase transparency and improve market competitiveness. Following the financial crisis of 2008, MiFID II and the corresponding MiFIR were discussed to introduce further legislation on the market. In essence, MiFID II comprises two levels of legislation, the directive itself (MiFID II) and a regulation (MiFIR). MiFID II entered into force in July 2014 and applied for all investment banks in the EU from January 2018. In addition, the framework legislation is supplemented by implementing measures ("Level 2 rules") which take the form of delegated acts and technical standards. These are still being introduced on a gradual basis and includes the introduction of selected ESG considerations in August of 2021.

In addition, the technological revolution, increased complexity of the financial markets and increased risk of financial crime, and the market abuse regulation (MAR) were introduced in 2016 to increase transparency. Replacing its predecessor (Market Abuse Directive [MAD]), MAR, like MiFID II, aims to protect investors by increasing transparency and limiting market abuse. MAR has a particular focus on information disclosure by companies—in particular financial reporting standards and commentary on material information.

Through the combination of MiFID II, MIFIR and MAR, the EU aims to offer investors better insight into products and services offered on the market, along with transparency in processes. Therefore, uniformed standards result in driving competitiveness in the financial markets. MiFID II mainly focuses on trading venues or structures in which financial instruments are traded. In contrast, MiFIR relates to regulating the operations of these venues and MAR concerns with market abuse. Even though the legislation is technically EU law, they are also concerned with third-country members (non-EU countries), applying worldwide.

The Framework of MiFID II/MiFIR

MiFID II has three core objectives dictated by the EU treaties:

1. Protection of investors and the integrity of the financial markets.
2. Promotion of fair, transparent, efficient and integrated financial markets.
3. Harmonisation of European trading and investment market.

These objectives are created through four main methods:

1. Increasing market transparency by higher focus on trading controls.
2. Framing investor protection with new organisational requirements for onboarding of investors.
3. Introduction to a market structure framework with clearing infrastructure.
4. Strengthening supervisory powers and effective harmonised sanctions.

From a practical perspective, these methods are included in a legal framework of three main topics that include a wide range of topics (Fig. 13.2).

How Does MiFID II/MiFIR Impact Listed Companies?

From reading the topics alone, it is clear how the legislation influences many aspects of the financial markets and has changed how they operate. This is especially true for investment banks, which have been forced to focus on compliance, but it has also changed the fundamental information flow between the participants of the financial markets.

Therefore, since the implementation of MiFID II, companies and their IR functions have changed how they work concerning corporate access for the buy-side and equity side. In the past, the sell-side would expedite and facilitate conversations between potential investors and the company. However, legal action has changed the landscape.

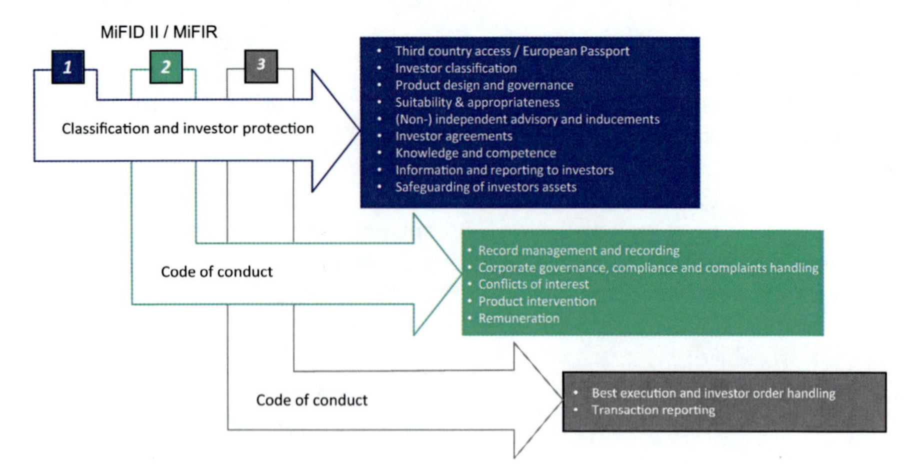

Fig. 13.2 Topics of MiFID II

Most topics are concrete actions that investment banks now need to consider when onboarding, communicating, trading and reporting to their current and prospective investors. However, three main topics are relevant for listed companies under MiFID II: (non-) independent advisory and inducements, investor classification and conflict of interest.

(Non-)independent Advisory and Inducements

In the previous two sections, we discussed how the financial markets are set up and the role of different financial markets participants in that framework. Investment banks attempt to create a value-added proposition to their investors by, for example, hiring expert equity analysts who advise major institutional investors on investment ideas. In the past, this research was freely distributed to buy-side investors to attract trading commissions hopefully.

MiFID II and its associated delegated directives (e.g. Chapter 4) specify that investment banks must distinguish between trading commissions, monetary benefits, non-monetary benefits and other fees. This innovation intends to promote transparency in the fee structure for investors, as services not associated with the actual trading of financial instruments must be separately accounted for and charged for.

Further specified (with operational requirements) in Article 13 of the EU's delegated directives, it is further specified that institutional investors are required to make direct ad hoc payments for research or set up a research account with the investment banks. This decision must be independent of the trading commissions. Therefore, the institutional investors are obligated to assess the quality of research and provide a budget and cost information to the investment banks and their stakeholders. Therefore, the profitability of the once-lucrative business of charging high trading commissions, while providing add-on research service, has been reduced.

As a result, each investor meeting, broker commentary or research report needs to be judged on a commercial perspective by the investment bank. Therefore, the price of trading has fallen significantly as institutional investors must opt-in for research, i.e. pay for it separately. In the meantime, investors have discovered that they did not need all the research that investment banks have historically provided, leading to a reduction in overall research fees.

In Article 12 of the EU's delegated directives, we find that the EU commission opens for the possibility of institutional investors receiving "minor non-monetary benefits", provided by the investment bank—essentially, a separation between "research" and "marketing". This includes:

- Information or documentation relating to a financial instrument, or an investment service, is generic or personalised to reflect the circumstances of an individual investor.
- Written material from a third party commissioned and paid for by a corporate issuer or potential issuer to promote a new issuance by the company. It can also be where the third-party firm is contractually engaged and paid by the issuer to produce such material on an ongoing basis. However, this is only provided that the relationship is disclosed in the material.
- Participation in conferences, seminars and other training events for the benefits and features of a specific financial instrument or an investment service. This is specified not to include corporate access (i.e. meetings between institutional investors and companies' senior management) facilitated by the investment bank (Fig. 13.3).

Investor Classification

As we illustrated in Part II, MiFID II distinguishes between three different classes of investors: retail investors, professional investors (either per-se or opt-up) and eligible counterparties (ECPs). One motivation behind MiFID II is that different investors' knowledge, experience, and financial abilities vary. They are therefore offered different levels of investor protection depending on their classification.

Whereas retail investors receive the highest protection (e.g. concerning information duties), ECPs are confronted with the lowest level of protection. However, investors can request a reclassification (i.e. opt-up or opt-down). Furthermore, the investor classification under MiFID II does not influence consumer protection laws (e.g. professional investors under MiFID II might still be classified as "retail" or "consumers" from a civil law perspective).

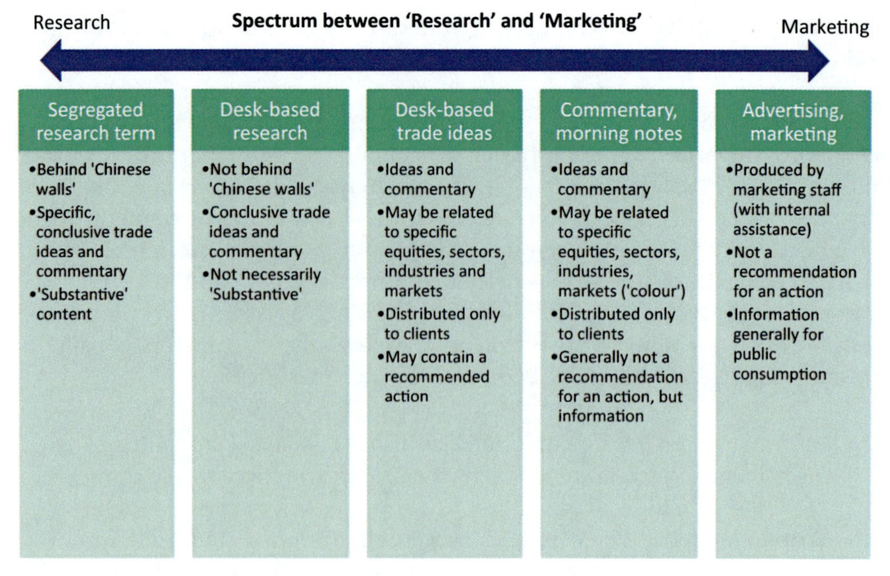

Fig. 13.3 Spectrum between marketing and research[5] (*Source* PwC [2016])

According to MiFID II, "retail investor" means every investor who is not professional. MiFID II distinguishes between per-se professional investors and opt-up investors.

1. **Per-se entities** authorised or regulated to operate in the financial markets. These entities include:
   - Credit institutions.
   - Investment banks.
   - Insurance companies.
   - Institutional asset management companies and funds.
   - Pension funds and their management companies.
   - Commodity and commodity derivatives dealers.
   - Locals (i.e. firms providing investment services that exclusively deal on their account in futures, options or derivatives for hedging positions).
2. **Per-se investors** that may also have large undertakings meeting two of the following requirements (irrespective of an authorisation to operate in the financial markets):
   - **Balance sheet total:** 20 million euro.
   - **Net turnover:** 40 million euro.
   - **Own funds:** 2 million euro.

3. **Per-se investors that are national and regional governments**, including public bodies that manage public debt at national or regional level (e.g. central banks, international and supranational institutions such as the World Bank, the IMF, the ECB and the EIB). The institutional investors whose main activity is to invest in financial instruments, including entities dedicated to the securitisation of assets or other financing transactions (e.g. certain family offices, corporate finance companies).
4. **Or opt-up professional investors** are, in principle, retail investors who may be treated as professional investors on request. However, several requirements must be satisfied to qualify as an opt-up professional investor.

ECPs are, in principle, investment banks, credit institutions, insurance companies, UCITS and their management companies, pension funds and their management companies, national governments and their corresponding offices, including public bodies dealing with public debt at the national level, central banks and supranational organisations. Therefore, ECPs are, in principle, provided with the lowest level of investor protection. However, MiFID II extended the protection of ECPs concerning the safeguarding of financial instruments and information and reporting requirements.

Investors other than professional investors, including public sector bodies, local public authorities, municipalities and private individual investors, may also be allowed to waive some of the protections afforded by the conduct of business rules. Therefore, investment banks shall be allowed to treat any of those investors as professionals provided the relevant criteria and procedure mentioned below are fulfilled.

However, those investors shall not be presumed to possess market knowledge and experience comparable to professionals. Any such waiver of the protection afforded by the standard conduct of business regime shall be considered valid if an adequate assessment of the investor's expertise, experience and knowledge is made. This is undertaken by the investment bank and gives reasonable assurance, considering the nature of the transactions or services envisaged, that the investor can make investment decisions and understand the risks involved.

- The requirement to implement appropriate written internal policies and procedures to categorise investors.
- Professional investors are responsible for keeping the investment bank informed about any change, affecting their current categorisation.
- Should the investment bank become aware that the investor no longer fulfils the initial conditions, which made the investor eligible for professional treatment, the investment bank shall take appropriate action.
- Impact on the Suitability & Appropriateness testing (Art. 54).

- The professional investor is deemed to have the necessary level of experience and knowledge to be financially able to bear any related investment risks consistent with the investment objectives.

- Limited application of information requirements (Art. 50).
- Only retail investors will be provided with a Statement of Suitability (Art. 54).
- Written basic agreements (Art. 58).

Investment banks that operate as sell-side brokers and corporate finance advisers are typically not very concerned with retail investors. They have strict compliant "Chinese walls" and a closed-loop business model where information flow is concentrated and confidential. However, in some circumstances, especially in recent years, they have adapted their business model to encapsulate retail investors, a topic we have briefly earlier commented on, and will revert to further.

Conflict of Interest

MiFID II requirements on the identification, prevention and management of conflicts of interest (CoI) are nothing new and largely like the requirements imposed by MiFID I. However, amendments on the degree of detail of the CoI procedures must be carried out. In particular, the following principles must be considered:

- The CoI policy, as well as the register/database of conflicts, must be expanded to include all potential risks, not only material risks (e.g. but conflicts also that can arise from remuneration and inducements structures).
- "Appropriate steps" regarding the identification and management of potential conflicts (measures which prevent or manage conflicts) must be undertaken.
- Disclosures of CoI to the investor is now a measure of "last resort".

The CoI procedure has quite specific requirements for the investment bank, namely:

- **Identification:** To identify, prevent or manage conflicts of interest, it could make sense to appoint the compliance function as conflicts of interest manager that is independent of the investor management section.
- **Prevention:** Ensure that your employees are fully aware of the CoI policy by providing training and education to the employees on organisational measures for the prevention.
- **Management:** Retain all the records concerning identifying and managing the conflicts of interest. Assess and monitor the adequacy and

effectiveness of the conflicts of interest management system regularly (at least annually).

- **Disclosure:** Disclosure to the investor is to be a "last resort". This means that disclosure can only be used where the bank's organisational/administrative measures are insufficient to ensure that risks of damage to the investor's interests will be prevented. In addition, procedures to disclose conflicts of interest must be established (e.g. standardised template).

Additional organisational requirements concerning investment research, recommendation or marketing communications must be considered. The requirement to identify, prevent and manage conflicts of interest applies to equity analysts or other relevant persons of the investment bank who produce or arrange to produce investment research or recommendation that is intended or likely to be subsequently disseminated to investors of the firm. An agreement concerning the behaviour of equity analysts must be designed following the principles:

- Personal transactions or trades in financial instruments to which investment research relates must not be undertaken.
- The physical separation must be ensured between equity analysts and other relevant persons whose interests may conflict with the interests of the person to whom the investment research or recommendation is disseminated.
- Inducements from those with interest in the subject matter must not be accepted.
- Favourable research coverage must not be promised to issuers and shall be completely independent by the equity analyst.

Investment banks that provide advice on corporate finance strategy and provide underwriting or placing financial instruments must also have CoI policies in place. Before accepting a mandate, the bank must provide the issuer investor with a specific list of information, including among others:

- The arrangements that are in place to prevent and manage conflicts.
- Job titles and departments of the individuals involved in the provision of investment advice.
- Various financing alternatives are available, and the number of transaction fees associated with each alternative.
- The investment bank must have in place a centralised process to identify all underwriting and placing operations and record such information, including the date on which the investment bank was informed of potential underwriting and placing operations.

- Ensure adequate controls are in place to manage any potential conflicts of interest between underwriting and placement as well as investment research and between their different investors receiving those services.
- If CoI cannot be managed by way of implemented measures, the investment bank must not engage in the operation.

The investment bank will have clear policies to handle potential CoIs; however, the company must realise what information is disclosed to which departments in the investment bank. If in doubt, it is best to disclose information to the corporate finance department, who can determine if the information is material and relevant and thus involve the equity analysts.

The corporate finance department can also include the equity analysts on the "insider list" (a continuously updated list of employees working at the investment bank that holds insider information submitted to the compliance department). Only once the information is publicly disclosed can the analyst again discuss the company with investors.

It is best practice for the corporate finance department to include all relevant brokers and analysts to the insider list when engaging in an ECM engagement. The reason is that during the engagement, an equity analyst may independently change their recommendation or view on a company which could cause problems in the process.

For example, corporate finance may be hired to raise capital for a company by issuing new shares. This process dilutes the shares of existing shareholders. Meanwhile, an equity analyst may downgrade a company from buy to sell, a potential clear benefit for the investment banks' investors. Of course, this is only done if corporate finance wins an engagement and is ready to announce the transaction publicly.

Embracing MAR

MAR is designed to deal with the strong reactions to market abuse and increase market attractiveness by enhancing investor protection. Expanding the definition of market abuse, MAR also aims to cope with the rapid growth of technology and the complications it brings. MAR is quite tangible and comprehensive and aims to standardise standards on the financial markets. Through the framework of MAR, companies have standards to identify, monitor and report market abuse and disclose insider information.

We discussed information on the financial markets in Part I. MAR disseminates information and defines market abuse as an investor missing information to gain an unfair advantage on the financial markets. MAR frames three main forms of market abuse:

- **Insider dealing:** This is the act of abusing inside information to make, amend or cancel deals or to encourage a third party to deal using this knowledge.
- **Unlawful disclosure of inside information**: This is the act regarding selectively disclosing inside information without sufficient and necessary grounds.
- **Market manipulation**: This is the umbrella term for a series of actions distorting market performance (liquidity) and pricing mechanisms for financial instruments.

Through the framework, the regulatory authorities in EU member states have a uniform approach to market manipulation and decentralised options for Competent National Authorities to enforce sanctions against non-compliant financial markets participants. It is important to note that MAR applies to financial markets participants, investment banks and investment banks, unlike MiFID II, which applies to financial instruments, behaviours and transactions. They are therefore concerned with:

- Financial instruments admitted to trading on a regulated market.
- Financial instruments traded on a multilateral trading facility (MTF), admitted to trading on an MTF, or for which a request for admission to trading on an MTF has been made.
- Financial instruments traded on an organised trading facility (OTF).
- Financial instruments not covered by point (a), (b) or (c), the price or value of which depends on or affects the price or value of a financial instrument referred to in those points, including, but not limited to, credit default swaps and contracts for difference.
- MAR also covers certain types of bids in terms of behaviours and transactions. MAR applies if the auction platform is considered a regulated market of emission allowances or similar auctioned products. This is still the case if the products being auctioned aren't considered financial instruments.

What Information Does a Company Need to Disclose Under MAR?

Companies need to disclose inside information as soon as possible to mitigate insider dealing and ensure an even playing field. According to MAR, inside information should be complete and free from errors. In addition, inside information should be disseminated to enable fast access to the public. From our experience, we have seen many examples of companies releasing non-material information to promote their company or delaying the disclosure of material information for the reason of postponing a negative share reaction. This is not MAR compliant.

Firstly, it is important to understand and determine if the information is material when disclosing inside information. Material information is information that a reasonable investor would consider important in making an investment decision. Therefore, the information needs to relate to a particular issue that can impact the company's earnings and business. Further, the company's IRO and senior management will also need to determine if the information may have a notable effect on the share price when the information is publicly released.

MAR specifies that the information should be outlined clearly and must not be in conjunction with marketing activities and must be publicly available on the website for five years. Information that the company deems non-material but wants to announce can be done so from a press statement rather than a company announcement. Buy-back programmes, stabilisation measures and accepted market practices are exempt from MAR when all other regulatory conditions have been met.

Delay of Disclosure of Inside Information

The delay of inside information is allowed if an issuer or emission allowance market participant (EAMP) sees fit. However, it comes with tight conditions (Art. 17). Delay is only permitted if:

- Immediate disclosure is likely to prejudice the issuer's legitimate interests.
- Delay of disclosure is not likely to mislead the public.
- The issuer can ensure the confidentiality of the information.

To stay compliant with MAR, all disclosing delays of inside information must be reported to local, national competent authorities immediately following disclosure. In addition, appropriate records should be kept outlining the details of the delay of disclosure, although these do not necessarily need to be submitted unless requested.

A company can release a company announcement on a particular risk or issue without specifying the details if a particular situation remains unclear. For example, during the COVID-19 pandemic, national shutdowns worldwide result in a volatile business environment. As a result, many companies released a preliminary company announcement stating that investors should know how the companies were currently affected. In addition, many companies also suspended their financial targets until they had a better overview of the situation.

Suspicious Transaction Reporting

Brokers must have an effective procure to report suspicious orders or attempted orders, or transactions. The procure must outline criteria for assessing suspicious activity and have surveillance systems to monitor and flag suspicious activity. Record-keeping, also the main topic in MiFID II, is essential for the compliance function to identify, track, file and communicate the reports to the competent national authorities. Under MAR, both reports are deemed suspicious and those flagged but not concluded unreasonable.

Announcing a Transaction: Market Soundings

As we earlier established, the main reason for a company to be listed on a stock exchange is to have liquidity in the shares and/or to facilitate later capital raises. This allows flexibility for the company's shareholder to trade their ownership essentially. An investor can trade shares on the open exchange. However, they can also enquire a broker to find to assist. For example, suppose a tier 1 institutional investor wants to buy or sell a significant number of shares. In that case, it can prove difficult and time-consuming if there is limited daily liquidity in the shares and can also lead to a large impact on the price of the shares. If a broker can find another institutional investor to take the opposite side of the transaction, they can transact a "block trade" (some exemption occurs) or "private placement" of the shares.

MAR includes the term "market sounding", which is exactly the act of a broker attempting to test the market and find an investor on the opposite side of the transaction. Therefore, the transaction is announced, and a broker will reach out to prospective investors and perhaps communicate the potential size, price and timing of an upcoming trade. In the circumstances where an insider of a company wants to sell a significant number of shares, for example, an early-stage private equity fund selling 20% of their shares following an IPO, the transaction is usually conducted after-market close not to distort the market price.

This procedure requires trust with its sensitive nature. Therefore, it is highly regulated by MAR (outlined in art. 11) and must always be digitally recorded for five years. Along with insider transactions, if a company were to consider an M&A situation, there may be a case to release this inside information to its shareholders. However, the conditions state that this is only allowable if the shareholders' opinion is crucial to deciding the merger.

If all protocols for a market sounding are met, disclosing market participants are covered against the suspicion of market abuse. They are granted with a "safe harbour" (exceptions where certain behaviours are considered exempt from MAR). If the applicable conditions are met, disclosing inside information in a market sounding is deemed a person's normal professional duty.

When a broker tests the market, the recipient needs to be informed that they will receive a market sounding, so the following information is insider information and is recorded. A broker will always call a list of usual counterparts and not involve unusual trading partners or cold calling. Therefore, the recipient will assess whether they are entitled to receive such information and declare this.

Investment Recommendations

Any investment research and recommendations made by a potential insider must be accompanied by a disclosure of interest. Due to the nature of the equity analysts' job, they receive information from multiple sources; it is, therefore, important to disclose the origin of the research. Therefore, the analyst's identity, job title, and relevant competent authority must be disclosed. Equally, disclosures should detail whether the recommendation is based on opinion or fact, and if so, which facts and any material assumptions accompany this. It should also include any forecasts, projections or price targets underlying the recommendation.

PDMR Transactions

It is considered material information if company's senior management is shareholders in the company they represent. It is material information if the company's senior management sells or buys shares. Therefore, under MAR, a Person Discharging Managerial Responsibilities (PDMR) is a member of the company's managerial, supervisory or administrative body. In addition, any senior manager with regular exposure to inside information and is in a managerial function to make key strategic decisions about the company.

The disclosure is for the insider to trade any financial instruments, including, but not limited to, shares, derivatives and debt instruments of the company. PDMRs must notify relevant authoritative bodies of all transactions in most countries once a predefined threshold has been reached within one calendar year. Depending on the member state, the threshold may be anywhere between 5,000 euro and 20,000 euro. Notifications of PDMR transactions must be made within three business days to stay compliant.

Record-Keeping and Insider Lists

Record-keeping is an integral element of MiFID II, MiFIR and MAR. One of the most important facets of record-keeping is creating and properly maintaining insider lists. Companies must create comprehensive lists that denote the names and identities of the persons within the organisation with access to insider information.

According to MAR and associated regulatory acts, these lists must be kept up to date and in the correct format. Entries on insider lists should include information about the date and time of receipt of inside information, along with extensive documentation of the individual's personal information at the time.

In this sense, entities are permitted to keep permanent and event-based insider lists. Permanent insider lists are reserved for persons who always have access to inside information. In contrast, event-based lists are for those who only receive inside information concerning certain events.

	MAR article Infringement	Maximum administrative pecuniary sanctions
"Natural" Persons	14 and 15	5,000,000 euros
	16 and 17	1,000,000 euros
	18, 19, and 20	500,000 euros
"Legal" Persons	14 and 15	15,000,000 euros or 15% of total annual turnover
	16 and 17	2,500,000 euros or 2% of total annual turnover
	18, 19, and 20	1,000,000 euros

Fig. 13.4 Sanctions of MAR infringement

Record-keeping is a stringent process for compliance with MAR. All suspicious reports, records of inside information, disclosure delays and so on must be retained for five years. Regulatory authorities have been granted the power to request these records at any time and can demand records of telephone conversations, data traffic records and electronic communication. Incorrect record-keeping is in breach of MAR and can lead to substantial financial sanctions, including:

- Add all insiders to a **permanent insider list**. If used, this list must be reserved for individuals who always have access to all inside information.
- Failure to create and update **event-based insider lists**. The purpose of this type of list is to enable authorities to determine what inside information exists within the company and at what time this information was identified.
- Not storing insider data in the **digital format** specified by ESMA in Article 18(9) of MAR.
- Failure to **notify insiders** and obtain their confirmation when including them on an insider list (Fig. 13.4).

References and Sources

1. MiFID II (2014/65/EU): European Union. (2014). On markets in financial instruments and amending 2002/92/EC and Directive 2011/61/EU. *Official Journal of the European Union.*
2. Financial industry regulatory authority (FINRA). (2010). *Rules and Guidelines*, www.finra.org/rules-guidance/rulebooks.
3. MiFIR (600/2014/EU): European Union. (2014). On markets in financial instruments and amending (EU) no. 648/2012. *Official Journal of the European Union.*
4. MAR (596/2014/EU): European Union. (2014). On market abuse (market abuse regulation) and repealing Directive 2003/6/EC of the European Parliament and of the Council and Commission Directives 2003/124/EC, 2003/125/EC, and 2004/72/EC. *Official Journal of the European Union.*

5. Source: PwC. (2016). The future of research: Impact of MiFID II on research for investment firms, https://www.pwc.com/gx/en/advisory-services/publications/assets/the-future-of-research-mifid-ii.pdf.

CHAPTER 14

Learning from the Impact on Financial Markets of Recent Legislation

Ensuring Fair and Transparent Markets

The past decade's regulatory wave has had profound effects on investment banks. They have faced decreasing income but also increased costs, effecting their historic very lucrative margins. We believe it is essential for the IRO to understand the situation of one of their most important stakeholders, i.e. the investment banks, to navigate how to collaborate with them.

In addition, we believe it is reasonable to assume that further legislation with new and amending requirements will continue to be introduced. The core objectives dictated by the EU treaties will be the same of protecting investors, promoting fair financial markets and harmonisation the investment market. Legislation introduced for the past decades has focused on defining the regulatory framework; in fact, MiFID II's motivation was to outline financial markets participants and increase transparency.

In the future, focus will be on defining innovative technological solutions and their associated regulatory implications, but also supplement new legislation to existing financial products and definitions. The implication, all else being equal, is that investment banks and companies will need to increase transparency through higher reporting standards, but also increase their focus on a "common good" approach. We believe it is wise learn from past effects of past and existing legislation and mitigation methods introduced in the market, before investigating the current new wave of regulatory standards.

P. Lykkesfeldt and L. L. Kjaergaard, *Investor Relations and ESG Reporting in a Regulatory Perspective*, https://doi.org/10.1007/978-3-031-05800-4_14

Regulatory Impact on Financial Markets

In 2017, Quinlan & Associates concluded from a large survey that the top 15 US investment banks published 40,000 research reports, where only 1% were read each year by investors.[1] From our experience, we believe the reading rate may be modestly higher in Europe. Historically, the low reading rate has not been an issue because research served as a free add-on to charge high execution fees when trading financial securities. Therefore, investors paid little attention to research payments, as they essentially received them for free, or more correctly, were part of the trading commission.

However, under MiFID II considerations of (non-) independent advisory and inducements, MiFID II banned the bundling of research with trade execution, compelling investment banks to price and sell their research as a separate product at fair and transparent market prices. Besides high transparency, this would also lower "overproduction of research". The investment mandate for all PMs is to act in the best interest of its clients. Therefore, the institutional investors are obligated to assess the quality of research and provide a budget and cost information to the investment banks—along with its stakeholders.

PMs also must disclose their budgets and costs to their investors, leaving them in a dilemma for research payments. They could choose to pass the cost on to their investors or absorb the cost in existing fees lowering their profit. Following the example of the largest asset managers in the world (BlackRock, Pimco, T. Rowe Price, Vanguard, etc.), most funds chose the latter and simultaneously cut external research costs (Fig. 14.1).[2]

The once-lucrative role of equity research has fundamentally been disrupted. In response to lower commercial potential of equity research, the budgets have been cut, leading to layoffs, shake-ups and cutbacks among equity research staff and lower remuneration. With lower budgets, investment banks have focused on providing equity research on larger listed companies, while specialised independent research firms have gained market share in the research coverage of listed SME companies.

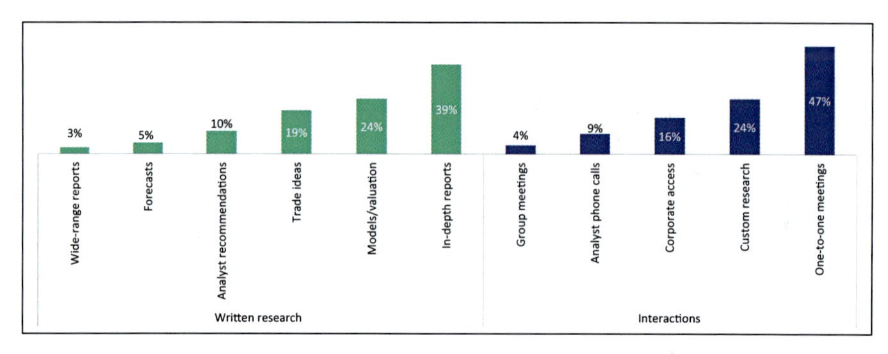

Fig. 14.1 How do institutional investors rate investment banks[3] (*Source* Bloomberg [2015])

The investment banks' earlier bundling of research gave investment banks and unfair advantage towards independent research providers, because the costs were cross-subsidised. MiFID II has therefore opened the market for specialised research agencies (independent of investment banks) with highly specialised knowledge on certain themes.

The disruption of the equity research industry and the supply of independent research firms have meant that both written research and equity analyst need to improve in terms of quality. It has been therefore important for the EU to lower "overproduction of research" where in the past equity research was used for marketing of company results rather than discovering an information advantage.

The disruption of sell-side equity research led, at least for a period, to turbulence and lower quality, and therefore, the largest tier 1 institutional investors have increased their internal research capabilities by hiring the best equity analysts.

Effects from Unbundling Research

Since the implementation of MiFID II in 2018, a vibrant debate concerning the effects on unbundling research has arisen. The effects seem to be quite clear, and we would like to highlight some of the main findings from significant research papers:

CFA Institute (2019).[4]

- Research providers have not benefitted from the new directive, as 57% of buy-side respondents report sourcing less research from investment banks than before MiFID II.
- The survey also presents mixed results regarding the quality of the overall research, as 48% of buy-side professionals believe that research quality is unchanged, whereas 44% of sell-side think that research quality has decreased.
- When the survey regards the perception of overall research coverage, 45% of buy-side and 52% of sell-side professional perceive a decrease in research coverage since MiFID II introduction.

The European Union (2020).[5]

- Almost all large asset managers chose to pay for research from their own P&L rather than by using research payment accounts. Survey and interview evidence suggest that the choice was largely influenced by competitive pressures which made it hard for asset managers to impose new explicit research charges on their investment clients.
- The average asset manager budget for external equity research fell by around 20–30% between 2017 and 2019. The decline was concentrated

in the budgets of larger firms (with the most sizeable firms reducing their research budgets by 30–40%). SME equity research expenditures also declined but not by as much as expenditures on large cap equity research.

- Many interviewees and survey respondents believe that MiFID II has had a negative impact on SME equity research. 71% of buy-side survey respondents reported that the effect of MiFID II on the availability of SME equity research was negative. While most of the interviewees view MiFID II as having a negative impact, quite a few, particularly from larger and more specialist firm, report that research volumes and coverage have not so far declined substantially.
- The coverage and research volumes did fall below trend in 2018 in several European regions and countries. The results indicate statistically significant negative 2018 effects (over and above trend) for most firm sizes in Western, Southern and Eastern Europe and for non-SME companies in Northern Europe. For SMEs, there is some evidence that the decline was partly cyclical and reflected low trading activity in 2018.
- Allowing for cyclical effects (using the statistical technique of instrumental variables), declines in SME coverage and research volumes are less severe in 2018 but, for mid- and large cap companies, declines in coverage and volume occur in 2018 that are statistically significant and about 10% in magnitude.
- Research quality (as measured by the standard deviation of Earnings per Share (EPS) forecast errors) did not change in a systematic way after the introduction of MiFID II.
- The statistical evidence on research coverage contrasts sharply with survey responses indicating a widespread industry view that research quality has declined. (Of buy-side respondents, 49, 44 and 7% said investment research quality fell, was unchanged or rose, respectively, since MiFID II. The buy-side respondents cited factors such as juniorisation, reduced analyst numbers and fewer research providers that led to the reduction in the quality of SME equity research.)
- The inducement restrictions introduced by MiFID II apply to other benefits that asset managers may obtain from brokers including the facilitation of contacts with issuers' top management that brokers have traditionally offered in the form of organising road shows and conferences at which issuers and asset managers can meet. These contacts, referred to by issuers as IROs, and by asset managers as Corporate Access (CA), can have an important impact on the financing terms that smaller firms, Mid-Caps and SMEs, receive in the market. Interview and survey participants suggested that IROs and CA for smaller firms have been affected by MiFID II. The access that issuers have to asset managers via brokers has become more challenging particularly when issuers want to elicit interest in foreign financial centres.

- Lastly, the report provides industry perspectives, as reported by survey participants, on how MiFID II has affected the liquidity of equity and corporate bond securities and on the costs of financing faced by issuers. 60% of issuers surveyed believe SME equity funding costs have risen and liquidity has fallen. It is noticeable, however, that individual issuers identify bigger implications of MiFID II for issuers in general than they perceive applying to their own firms (only 25% say their firm's access to equity or bond finance had worsened between 2017 and 2019).

Anselmi, G. and Petrella, G (2021).[6]

- Clear indications of a decrease in analyst coverage as well as a worsening in financial markets liquidity.
- SME and large cap company suffer a reduction in analyst coverage whereas liquidity drops only for SME shares, as the bid-ask spread for large caps is unaffected from the introduction of the directive.
- The finding of mild but steady signals of an overall worsening in price efficiency. This is the effects on market liquidity and price efficiency.

Effects on Corporate Governance

Unbundling of research has reduced the income of the sell-side investment banks. In addition to the technological revolution that has decreased the cost of trading and the challenges the buy-side has with the rise of passive investments, the industry faces income challenges.

However, along with the income challenges, costs have also increased. In 2018, all financial institutions spend 240 billion euro annually on compliance and regulatory costs and have paid out 276 billion euro in fines since the 2008 finance crisis. It is estimated that today 10–15% of staff work with legislation, compliance, risk management and corporate governance. A figure that has risen dramatically in the past decade in line of the regulatory wave and is expected to continue.[7]

The increased requirements on regulatory standards have amplified the need internal compliance, procedures, reporting standards and ethics considerations. Legislation requires investment banks to have an increased focus on know-your-client (KYC) policies to make sure that there are properly classified and receive appropriate information in relation to their classification.

CoI, suspicious reporting, market sounding and overall reporting standards also require a considerable back-office to administer the daily activities. Front-office employees are now required to participate in regular workshops and complete tests to adhere to the most recent legislation.

We are not able to fully judge the net positive or negative of MiFID II. However, the above findings coincide well with our experience on the disruption on the occupation of equity research. Therefore, the IRO, especially as a representative of an SME, must consider the changing role of the equity analyst and investment banks, and be more considerate of the declining business case from analyst coverage of companies. Chapter 15, we explore ways to navigate this disruption.

References and Sources

1. Quinlan & associates. (2017). The rise of online research marketplaces. *Research.com.*
2. Mooney. A. (2017). Majority of asset managers to absorb external research costs. *Financial Times.*
3. Bloomberg. (2015). Broker vote survey, https://www.bloomberg.com/professional/blog/future-investment-research-post-mifid-ii/
4. CFA Institute. (2019). MiFID II: One year on, Assessing the market for investment research. *CFA institute.*
5. European Commission. (2020). The impact of MiFID II rules on SME and fixed income investment research: final report. *Directorate-General for Financial Stability, Financial Services and Capital Markets Union.*
6. Anselmi, G., & Petrella, G. (2021). Regulation and stock market quality: The impact of MiFID II provision on research unbundling. *International Review of Financial Analysis.* Vol. 76.
7. KPMG. (2018). *There's a revolution coming: Embracing the challenge of RegTec 3.0.*

CHAPTER 15

How to Optimise IR Within the Existing Legal Framework

Mitigating the Sell-Side Disruptions

As established, the consumption of buy-side research consumption has decreased. Yet, the industry seems to have mitigated the effects in large cap companies by reducing their attention to SMEs. Therefore, the clear losers of MiFID II are the SME companies that have realised a worsening price efficiency, higher bid-ask spreads, lower research attention and higher volatility.

We previously mentioned that in Article 12 of MiFID II, the EU commission introduced the term "minor non-monetary benefits", provided by the investment bank. This means that the investment bank can assist in marketing of a company towards investors but must disclose this relationship to its investors.

In addition, MiFID II is concrete on investor classification that the investment bank must have a clear onboarding procedure of its clients and need to make sure that they receive information appropriate to their classification. Marketing is not subject to investor classification, because it is not deemed investment research. This allows the investment bank to be in a unique situation because it can leverage its brand and quality of its analysts to publish marketing material on behalf of a company to both its onboarded professional clients and the general public. Particularly for SMEs, the IRO has three methods to mitigating the adverse effects of MiFID II: commission research, commissioned corporate access and digital IR which typically also come at a cost.

P. Lykkesfeldt and L. L. Kjaergaard, *Investor Relations and ESG Reporting in a Regulatory Perspective*, https://doi.org/10.1007/978-3-031-05800-4_15

Commissioned Research

The equity analysts at banks are deemed independent and must not be influenced by the brokers, corporate finance or other functions of the banks. Therefore, they are not allowed to publish positive equity research for any other reason than the analyst's own opinion. A company's relationship with the investment banks' other departments, e.g. credit, corporate finance or risk management, cannot be taken into consideration.

For this reason, the equity analysts typically have a stronger relationship with the institutional investors, as opposed to brokers who are compensated through trading commissions and corporate finance who are compensated from raising money from the institutional investors. Equity analysts are compensated more greatly from the investor's perception about them.

Following an in-depth research process, the equity analysts will publish a report with a trading recommendation and a target price. The report is distributed only to professional clients (that are onboarded), who can consider the report but, in the end, make their own decisions. Therefore, a subjective report with a recommendation and a target price is defined as equity research.

Suppose the report was written objectively with no recommendation. The report may include a fair value range (as opposed to a target price), with a positive and negative price range based on disclosed assumptions about the future of the company. All else equal, the report is very similar to a research report. This is considered possible and is deemed as "minor non-monetary benefits". The report can therefore be read by anyone, as it is considered marketing. The report can be paid for by a company to increase its marketing towards institutional investors.

The pros of commissioned research:

- It is a platform for SMEs to increase their exposure towards institutional investors.
- The analyst report can be distributed widely to retail investors as it is considered marketing, not equity research.
- The company can distribute the report to potential investors, so that they do not have to run through the business model from A to Z with every new investor.
- Like the relationship with an equity analyst that conducts regular research on a company, a company has a partner in the financial markets who can pass on information from investors and other market participants.
- As the equity analyst is objective, the company can receive more transparent information from the financial markets, as opposed to an equity analyst talking their own case.
- The investment banks which offer commissioned research see the income as steady and may therefore be positively biased versus the company.

The cons of commissioned research:

- Independent research is free for the company, and commissioned research has a cost. Depending on the size of the investment bank and scope of activity, we estimate the total cost between 25 and 60,000 euro annually, however, naturally with exceptions.
- As a commissioned research report is deemed as marketing, some investors, especially major institutional investors with buy-side analysts, see little or no use for the report.
- The function of equity analyst is highly associated with neutrality and subjective research. Writing commissioned research can therefore negatively affect the standing of an equity analyst towards institutional investors.
- If a company buys commissioned research from one investment banks, the likelihood of another investment bank providing independent research will often be reduced.

However, it should also be mentioned that some large cap companies, who already enjoy a significant amount of independent research coverage, decide to engage perhaps one firm or investment bank to produce commissioned research with the sole purpose of creating an increased dialogue with retail investors as the company may then make the commissioned research available on its website.

Commissioned Corporate Access

The investment bank is active in setting up investor events and presentations. Like equity research, SMEs have experienced a fall in CA events hosted by investment banks. This is especially true if the company is seeking to meet with potential investors in foreign financial centres, different from where its shares are listed. A company can pay an investment bank to setup meetings or invite their list of investors to physical or digital presentations.

Typically, commissioned CA is tied together with commissioned research. If a company pays for analyst reports, then it is in the interest for the investment bank to introduce the company to investors on an ongoing basis. This keeps the commissioned research intact and has strong synergies with the analyst work on writing reports. The pros and cons are in essence like commissioned research.

Independent Digital IR

Instead of paying say 25,000–60,000 euro for a commissioned relationship with an investment bank, some SMEs, often those with a B2C focus, can choose to leverage their social media presence and combine their marketing of products/services with their marketing of their stock market listing. In connection with digital IR, the company typically either creates its own social media platform or website, social media groups, or uses the platform of a digital IR provider who will often charge either a monthly fee, or a fee per event. Digital IR activities are currently predominantly aimed at retailed investors.

Digital IR is relatively new and untested; however, we believe this is a constructive and positive approach to increase the company's exposure towards retail investors. The company can also engage with social media groups and forums that discuss investment ideas. It is clear that the digital IR strategy need to be leveraged through existing marketing methods; otherwise, it may prove time-consuming. The company can also produce and publish small video clips following the announcement of the company's financial results or significant news. This should be seen as a supplement to the regulatory standards of reporting but can increase attention for the company.

For a B2C company, the company may also employ a "shareholder benefit programme" towards retail investors, subject to possible legislation in the area. The programme may have strong synergies with the communication and marketing department, as retail investors also act as "brand ambassadors" and customers for the company. It is important to seek advise from legal counsel as in some jurisdictions such an arrangement may be considered a hidden taxable dividend for the shareholder, i.e. the receiver of the dividend.

It can also be mentioned that, just as in the case of commissioned research, some large cap companies have also embraced digital IR with a view to get close to, and interact with, the retail investor community—acknowledging the increased importance and influence of retail investors.

CHAPTER 16

The New Wave of ESG Regulatory Framework

An Apparent Increase in Sustainability

Considering the technological revolution and global trade, the world is becoming more interconnected through information, resources, people and ideas. The narrowness of only thinking about the outcome of one's actions is diminishing, as it is clearer how actions have ripple effects on others. We are seeing this mindset spread out at an increasing rate in, e.g. companies, investors, investment banks and government institutions.

Through corporate social responsibility (CSR), most companies and investors have already embraced sustainability, and increasingly, this megatrend will underpin all future legislation. However, even though commitments are high, and sustainability is included in all newly announced strategy updates from companies, it seems that few have sincerely considered or implemented the reporting and regulatory aspects of sustainability.

As a call to companies to align their strategies and operations with universal principles related to human rights, labour rights, the environment and anti-corruption, the United Nations (UN) former secretary general, Kofi Annan, set up the Global Compact (UNGC) in 1999. Today, the organisation has developed into the world's largest sustainable business initiative, where more than 13,000 companies and 3,000 organisations together strive to anchor universal principles and Sustainable Development Goals (SDGs) into the heart of business strategies, and report annually on their progress to the UN secretary general.

Derived from the UN conventions and declarations, the members sign-up to align their business actions with ten universal standards considering human rights, labour, environment and anti-corruption. Since it was founded, the UN

P. Lykkesfeldt and L. L. Kjaergaard, *Investor Relations and ESG Reporting in a Regulatory Perspective*, https://doi.org/10.1007/978-3-031-05800-4_16

has updated its goals, which the UNGC has embraced and urged the private sector to contemplate.

Replacing the UN's millennium development goals (MDGs, 2000–2015) that focused on government intervention; the 17 SDGs (and 169 underlying targets) were introduced in 2015 emphasising on business participation in creating a better world.

Outlining non-binding targets and guidelines for companies to alleviate social, economic and environmental problems is meant to create global awareness. All UN member states unanimously agreed and joined together in defining the SDGs to create a 2030 plan for people, prosperity and the planet. The 17 SDGs are now the basis on a worldwide scale for implementing non-financial/ESG reporting initiatives and ESG legislation.

Depending on a company's exposure, it is recommended for any company to consider the SDGs in their reporting and outline which targets are most relevant for their business. Therefore, it is highly recommended for any company to tailor their non-financial/ESG reporting with their underlying business. In addition, not all SDGs are applicable for all companies, so it is recommended to select and deep dive into the most relevant SDGs for the company in question. Already in 2017 (not updated since), 85% of all Nasdaq-listed companies had chosen which SDGs to focus on—for example, the global stock exchange company, Nasdaq, has placed particular emphasis on SDG 5 (gender equality), 12 (responsible consumption and production), 13 (climate action) and 17 (partnerships for the goals).

The EU's Green Deal

In December 2019, the EU presented its green deal framework. It was the beginning of a long journey to implement legislation and a policy response to contribute to a carbon-natural EU society by 2050. In the wake of the COVID-19 crisis, the green deal adapted to include the recovery and resilience plans to stabilise the EU economy. The green value has three main objectives:

- Invest and mobilise ~1 trillion euro (1.1 trillion US dollar) to support sustainable investments between 2021 and 2027 to ensure a green transition.
- Create a framework for investors and the public sector to facilitate sustainable investments.
- Provide support to public administrators and project promoters in identifying, structuring and executing sustainable projects (Fig. 16.1).

As a result, a key objective of the European Commission's plan is to reorient capital flows towards sustainable investment and ensure market transparency. To achieve this, EU's taxonomy was created, a classification system for sustainable activities. Taxonomy is a robust and science-based tool for companies

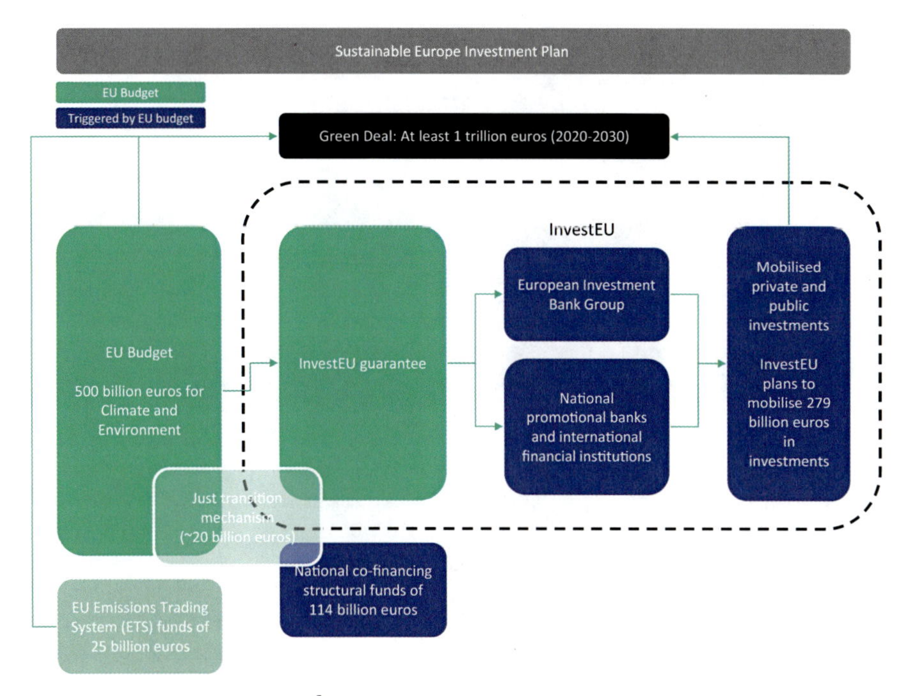

Fig. 16.1 EU's Green deal[1] (*Source* Eurocoop [2020])

and investors that provides criteria for determining which economic activities substantially contribute to the Green Deal objectives.

With the EU's green deal framework based on the UN's SDGs, the EU in 2019 published six thematic clusters, considered as the future drivers for change.[2] The SDGs and these clusters are meant to be underlying themes that will assist to shape the green deal's overall environmental and sustainability goals towards legislation on companies. Besides the overall EU treaties (Chapter 13), it is evident, or at least widely expected, that the EU will include these thematic clusters directly and indirectly in most, if not all, future legislation on the financial markets (Fig. 16.2).

Major Legislation on Sustainability in the EU

The EU has already announced and outlined major legislation to be implemented in the coming years. In the financial markets, the vast majority is tied to sustainability. This will likely have high impact on the flows of capital into certain shares and sectors depending on their standing compared to the relevant legislation.

Already today, an increasing number of investment mandates in funds require companies to adhere and comply with certain regulatory standards, including ESG, as an investment factor. The EU is clear about wishing to push

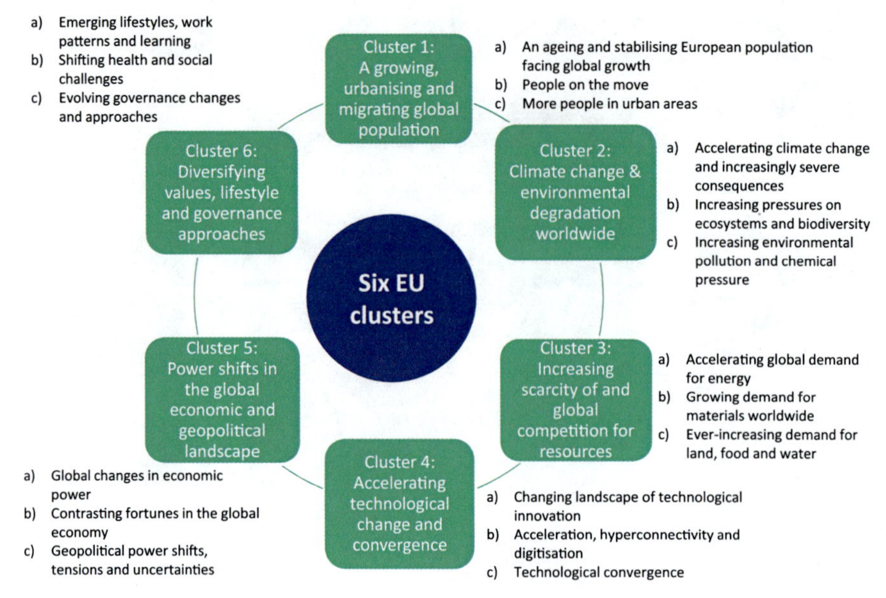

Fig. 16.2 EU/EEA six thematic clusters[3] (*Source* EEA [2022])

investments towards the sustainability area as part of the green deal. There are three major regulatory standards currently being implemented, which we further explore in Parts XI–X:

- An overall plan and definitions of sustainable activities outlined in EU's taxonomy delegated acts on sustainable activities.[4]
- A directive for companies (non-financial companies) to implement taxonomy: EU's Corporate Sustainability Reporting Directive (CSRD) (prev. non-financial reporting directive (NFRD)).[5]
- A regulation for investors (financial companies) to implement taxonomy: EU's Sustainable finance Disclosure Regulation (SFDR).[6] Amendments to MiFID II were introduced in August of 2021,[7,8] have largely embraced and aligned the directive with the SFDR. However, they have broadened the number of applicable investment firms (Figs. 16.3 and 16.4).

The EU's Taxonomy

The EU's taxonomy for sustainable activities aims to assist the EU in scaling up on sustainable investment and implement the European green deal. It thus tries to establish a science-based dictionary defining what is (hopefully unambiguously) sustainable and sets disclosures requirements for companies and investment companies.

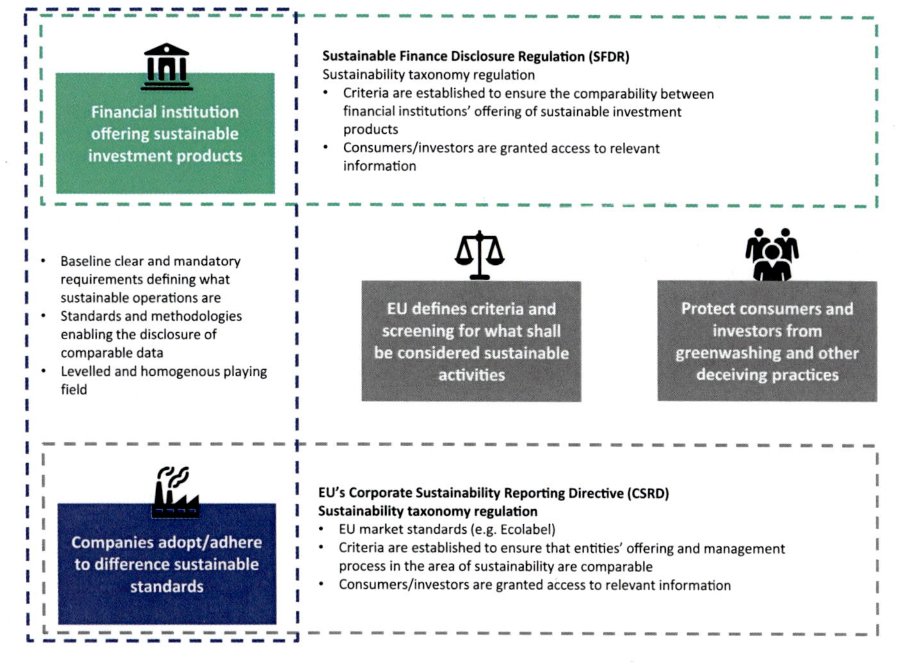

Fig. 16.3 An overview of taxonomy, CSRD and SFDR[9] (*Source* Boffo, R. and Patalano, R. [2020])

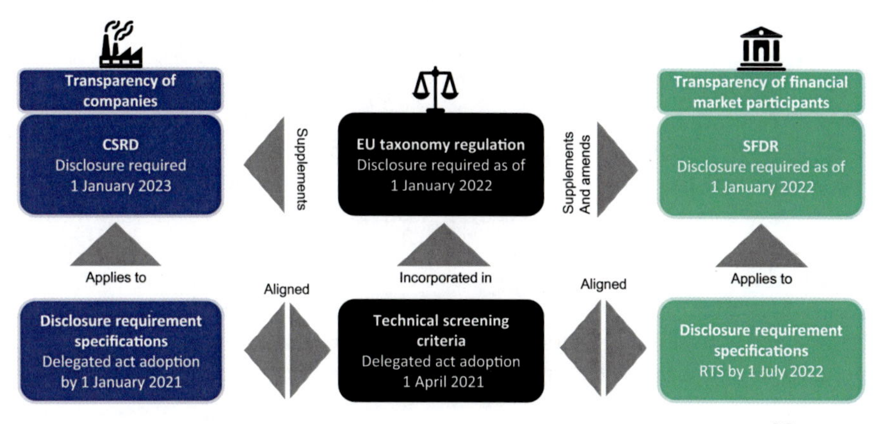

Fig. 16.4 Interconnected legal framework of taxonomy, CSRD and SFDR[10] (*Source* Verhey, M. et al. [2021])

It is essentially a green classification system of a list of economic activities (across 70 sectors) based on scientific and industry experience that can be adapted to changes in the marketplace, new activities and technologies. Taxonomy provides a framework for high emitting sectors for them to determine under what conditions they can be considered as environmentally sustainable. According to the European Commission, their key objective within the EU sustainability strategy is to:

> EU's efforts have predominantly focused on supporting investment flows towards economic activities that are already environmentally sustainable and towards plans to make them environmentally sustainable. A more supportive framework is needed to address the challenge of financing interim steps in the urgent transition of activities towards the EU's climate neutrality and environmental objectives.

In the past, companies considered CSR and making donations as best-practice principles. Today, ESG alignments must be integrated into the company's supply chain—from making and consuming products to its CAPEX. Sustainable supply chain strategy and non-financial reporting on it is important from a regulatory perspective to onboard all relevant companies in the taxonomy framework.

Therefore, the company must report their scope 1, 2 and 3 carbon emission footprints based on their direct and indirect consumption to also pressure its own strategy and processes to source sustainably. The legislation is a fully embedded framework for companies to integrate into their supply chains and aims to establish a common language to investors of what activities have a substantial impact on the climate and the environment.

Scope 1 and 2 activity concerning the company's own emission and energy supply is manageable; however, companies will find it especially difficult to implement scope 3 concerning its external emissions (Fig. 16.5).

EU's taxonomy is an overarching EU framework of sustainable investment guidelines, definitions and classifications that defines when an activity can be considered sustainable from an environmental perspective. In connection with taxonomy, there have so far been introduced over 50 directives, regulations and delegated acts (e.g. eco-design directive, EU water legislation, EU waste list, etc.) under the taxonomy umbrella addressing outlining the detailed framework.

Notably, the companies that are compliant with the CSRD (see next section) are required to include taxonomy in their non-financial reporting (art. 8). An overview of the elements of taxonomy is provided here (Fig. 16.6).

CSRD (Previously NFRD)

Implementing the standards of taxonomy, the intention is to limit "green-washing", by introducing harmonised reporting principles and thus increasing transparency. As a result, the mobilised resources from the green deal can be more efficiently invested to ensure the green transition in Europe.

In the light of taxonomy and the green deal, the EU commission reviewed the Non-Financial Reporting Directive (NFRD) as part of the strategy to

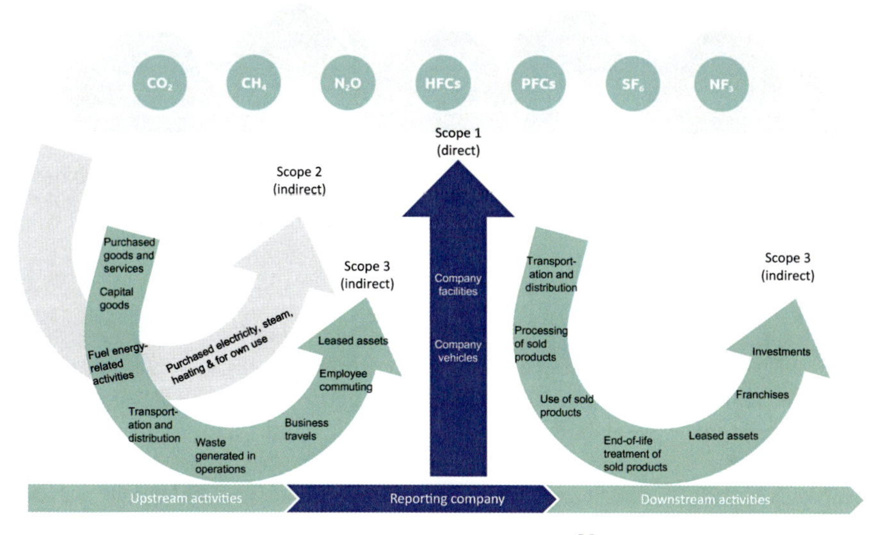

Fig. 16.5 Scope 1–3 of taxonomy in CO_2 emissions[11] (*Source* Green House Gas Protocol [2014])

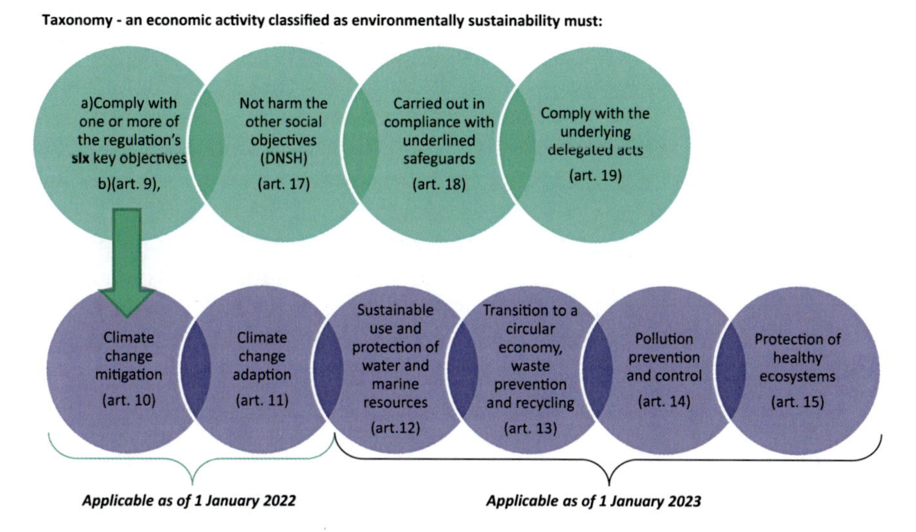

Fig. 16.6 An overview of the taxonomy legislation

strengthen the foundations for sustainable reporting. The NFRD will be re-branded to CSRD from 2023, aims to align with taxonomy and increases corporate disclosure requirements. All EU countries have already included the NFRD and the member states' laws and have enacted legislation regarding non-financial reporting.

Country	Local law	Country	Local law	Country	Local law
Austria	Act. 257 / ME of 6 December 2016	France	Art. L225-102-1 of 19 July 2017	Malta	Act CAP 386 of 12 February 2016
Belgium	Amd. 2564 of 3 September 2017	Germany	CSR-Rich.-Umsetzug. of 11 April 2017	Netherlands	L330 14 March 2017
Bulgaria	Amd. 237 of 6 June 2017	Greece	Law 4403/2016 of 07 July 2016	Poland	Amd. Act 61 of 15 December 2016
Croatia	Amd. 51 of 12 December 2016	Hungary	Amd. Act C of 2000 on Accounting of 15 May 2016	Portugal	D. Lei n. 89/2017 of 21 August 2017
Cyprus	Amd. 3 of 19 May 2017	Ireland	S.I, 360 of 30 July 2017	Romania	Ord. 1.938 of 17 August 2016
Czech Rep.	Amd. 462 of 14 December 2016	Italy	D.Lgs. 254/2016 of 30 December 2016	Slovakia	Amd. Act 130/2015 of 6 May 2015
Denmark	Stat. Act Amd. of 21 May 2015	Latvia	Law 2016/254.4 of 29 December 2016	Slovenia	Amd. ZGD-1J of 29 March 2017
Estonia	Acc. Act Amd. of 10 December 2015	Lithuania	Amd. XIII-94 of 15 December 2016	Spain	R.D. Ley 11/2018 of 28 December 2018
Finland	Amd. 1376/2016 and 1441/2016	Luxembourg	Law A156 of 29 July 2016	Sweden	2016/17"CU2 of 25 October 2016

Fig. 16.7 CSRD regulatory implementation in EU member states[12] (*Source* GRI [2017])

The companies included under CSRD are classified as listed and non-listed companies that have more than 500 employees, or either have a balance sheet of over 20 million euro or net sales of 40 million euro. There are slight country variations; for example, Denmark, Sweden and Iceland include large companies as having a minimum of 250 employees. There are an estimated 6,000 companies fitting these criteria and that must publish a sustainability report as of 1 January 2023. In addition, it is proposed that SMEs must also publish a sustainability report as of 1 January 2026.

According to CSRD, a company must include the proportion of taxonomy classified business that is included in their business activity. This includes their revenue, OPEX and project investments measured in CAPEX.

It is important to separate a company's business activity in this regard. For example, a company may derive 80% of its revenue from sustainable products ("taxonomy eligible") that contribute to a green society. However, it may have considerable non-sustainable operating costs associated with producing and selling the products (e.g. coal-powered factory plans and a fleet of gasoline cars) and plan to make non-sustainable investments in the future.

Separating activity increases transparency on the company's operations. The principal aim of CSRD is to enable the financial community, consumers and other stakeholders to evaluate the non-financial performance of companies, and to encourage those companies to develop a more responsible approach to business (Fig. 16.7).

SFDR

Almost 50% of the green deal's 1 trillion euro sustainable investment plan is based on mobilising private and public investments. Therefore, it is important to increase non-financial reporting standards in the investment industry—otherwise, it would not be possible to earmark sustainable investing.

SFDR aims to provide more transparency on sustainability within the financial markets in a standardised way, thus preventing greenwashing and ensuring comparability. Today, an EU investment company is required to disclose their primary holdings included in their fund. Now, the financial company needs to incorporate taxonomy and CSRD into their investor prospect material.

Like how a company may separate taxonomy-eligible activity, a fund can weigh the taxonomy eligibility of each company in the portfolio relative to the size of the position. The aggregated weight along with sustainability policies and disclosures needs to be disclosed and updated on a fund level.

The change in reporting requirements aims to get investment managers more aligned in non-financial reporting, including the CSRD disclosure of each company in their portfolio. If this is not aligned, then an investment company will be forced to liquidate its positions based on non-financial disclosures (as opposed to the equity analysis).

The outcome is an SFDR ranking based on a sustainability focus, which will be reported to the financial authorities in various categories and for marketing purposes. SFDR is already applicable from 10 March 2021 for most investment companies and is likely to become mandatory for all investment companies by 1 January 2025.

In August of 2021, the EU introduced adjustments to their amendments of MiFID II which are applicable from August 2022. Notably, these adjustments integrated the requirements of SFDR under the MiFID II framework and thus widened the scope of applicable financial firms. Essentially, under the amendment, all investment firms must disclose their ESG approach and exposure, as well as assure that their products and advisory services are aligned with their clients' preferences.

Based on the above, we believe that a number of companies and financial institutions may experience initial challenges in the non-financial reporting area.

A summary of facts and best practices is illustrated on page 329.

References and Sources

1. Eurocoop. (2020). Green deal and farm to fork strategy, https://www.eurocoop.coop/news/282-DIGEST-Green-Deal-and-Farm-to-Fork-Strategy.html.
2. European environment agency (EEA). (2019). *Drivers of change of relevance or Europe's environment and sustainability.* ISSN 1977–8849.
3. EEA. (2022). Drivers of change: Challenges and opportunities for sustainability in Europe. *European environmental agency.*
4. Taxonomy (2020/852): European Union. (2020). On the establishment of a framework to facilities sustainable investments and amending regulation (EU) 2019/2088. *Official Journal of the European Union.*
5. CSRD (2014/95/EU): European Union. (2014). Amending directive 2013/24/EU, amending Directive 2013/34/EU, Directive 2004/109/EC, Directive 2006/43/EC and Regulation (EU) No 537/2014, as regards corporate sustainability reporting. *Official Journal of the European Union.*

6. SFDR (EU 2019/2088): European Union. (2019). On sustainability-related disclosures in the financial services sector. *Official Journal of the European Union.*
7. MiFID II delegated acts (C(2021/2615)): European Union (2021). Changes to regulation (EU/2017/575) as regards the integration of sustainability factors, risks and preferences into certain organisational requirements and operating conditions for investments firms. *Official Journal of the European Union.*
8. MiFID II delegated acts (pending legal number at the time of writing this book): European Union (2021). Changes to regulation (EU/2017/593) as regards the integration of sustainability factors and preferences into the product governance obligations. *Official Journal of the European Union.*
9. Boffo, R., & Patalano, R. (2020). *ESG investing: Practices, progress and challenges*. OECD Paris.
10. Verhey, M., et al. (2021). *Making sense of the EU sustainable finance regulations*. Ramboll.
11. Greenhouse Gas Protocol. (2014). *GHG protocol agriculture guidance: Interpreting the corporate accounting and reporting standard for the agriculture sector*. Greenhouse Gas Protocol.
12. GRI. (2017). Member State Implementation of Directive 2014/95/EU. *Accountancy Europe: Policy and reporting.*

PART IV

Achieving a Fair Valuation of the Company Through Best-Practice IR

CHAPTER 17

The IR Function

Communication works for those who work at it.

—John Powell, Film-Composer

Introduction to Best-Practice IR

IR is an essential part of a listed company's integrated internal and external communication strategy. IR is not only about marketing a company towards investors, but it is about strategically managing relationships with internal and external stakeholders in need of information. The interaction between these communication areas is touched on in this chapter, but the focus is on IR. It is also important to note that even though a company has a specified IRO, best-practice IR must be performed on a listed company's senior and managerial levels. Therefore, we do not confine the term IRO, but any relation management has with the financial community and non-financial stakeholders.

In our experience, we have come across the notion that there is no information vacuum in the financial markets. A lack of information exacerbates and increases the severity of a crisis and may induce rumour or gossip. If the IRO does not provide the financial community with a constant flow of information, this can cause issues on the company's reputation.

Therefore, IR's primary role is to provide the financial community with the necessary tools and information to fairly price the company's shares (and bonds). It is not the function of the IRO to achieve the highest absolute share price but the fairest share price relative to the intrinsic value. Naturally, this should be evident—however, not always. It is not uncommon for some IROs, or perhaps more likely senior management, to complain about the share price (well, that is when the share price appears on the lower side)—which is never

P. Lykkesfeldt and L. L. Kjaergaard, *Investor Relations and ESG Reporting in a Regulatory Perspective*, https://doi.org/10.1007/978-3-031-05800-4_17

beneficial. The rise during the past decades of increased management warrant incentive programmes does on a regular basis result in the management being too focused on the share price. History is shown multiple examples of this leaving to a value-destructive short-termism. Companies should be aware that in essence institutional investors prefer that senior management concentrates to running the business in an efficient manner, and leave for the financial markets to focus on the share price and determine its correct level.

Senior management, board of directors or designated IRO complaining about the company's share price is not a sound strategy because it tells the market one of two things: either the aggregated collection of buyers or sellers in the financial markets are collectively inadequate. Therefore, the market dynamics are broken concerning the company's valuation. Alternatively, the company's representatives are unaware that they have failed to communicate its specifications and strategy to provide appropriate information to the financial community (Fig. 17.1).

Like strategic communication, IR is also strategic. This means that the function underpins a company's strategic objectives and milestones. These objectives are, namely, associated with the financial flexibility to explore opportunities in their capital structure, e.g. reducing the long-term risk premium. Therefore, a company's IRO commands the closest collaborative role to the financial markets. Furthermore, as the risk premium is also associated with the fundamentals of the businesses that are becoming increasingly intertwined with ESG, stakeholder management is becoming an increasingly important part of performing best-practice IR.

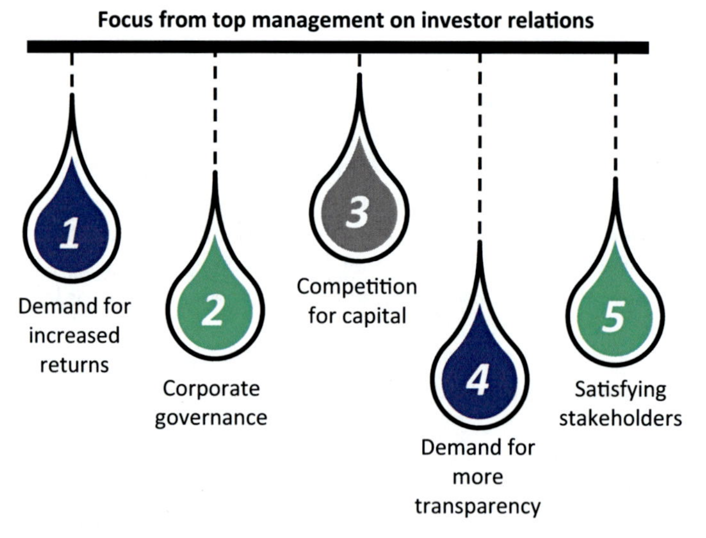

Fig. 17.1 IR as a core competence

We have previously asserted that investors make capital available for the company that can best employ the capital at a given time. Therefore, the PM seeks the highest return on the investments by making the capital available. In the neoclassical sense, a company is comprised of labour and capital. Consequently, labour is trading work hours for a wage, which is an opportunity cost relative to spending time elsewhere. Likewise, investors also trade capital availability as an opportunity cost to other investments with the expectation of gaining a return on the invested capital. Therefore, the HR and IR dynamics are vital for effective business practices.

IR is well established and has been a core competency for decades, especially in the US and the United Kingdom, partly rooted in Anglo-Saxon investment culture. This interest is partly a function of the increasing competition between the various portfolio management companies and investment associations and has led to increasing demands on companies for increased returns. Other factors have been increased demands for openness to external stakeholders, competition for funding and developments in the field of corporate governance.

Objectives of an IRO

It is important to emphasise that the IR department should be a sober and balanced information provider and promote its shares uninhibitedly. For example, if the company's share price falls out of line with the company's intrinsic value. In that case, the fall in the share price will only become more significant on the day the share may be re-rated by the financial markets in a downward direction. This means that the financial markets change its view of the shares, typically due to one or more equity analysts' changes in a recommendation (Fig. 17.2).

There is no short and precise definition of IR, but the purpose of IR typically includes the objective of:

- To secure and maintain the company's financial strength and flexibility in the short and long term, including minimising the costs of equity and debt financing.
- A stable price development (adjusted for the overall market development) on the company's shares, which realistically reflects the company's fair value.
- Good liquidity in the company's shares, e.g. by way of increasing analyst coverage.
- Flexible financing of any acquisitions by issuing new share capital, as the company's shares are considered an attractive means of payment.
- Act as a good business partner to the board of directors, senior management and BU managers.
- Minimise the probability of a takeover attempt (as the company's share price should reflect the company's fair value best possible).

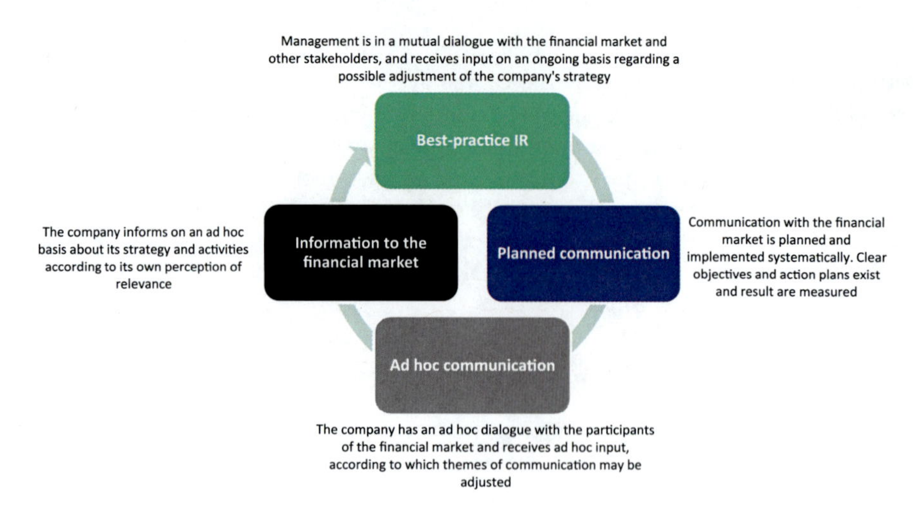

Fig. 17.2 Best-practice IR

- The company complies with the legislation and rules associated with being a listed company.
- The financial and non-financial standards are updated with the newest trends and legislations.

These objectives are sought to be achieved, among other things, by:

- First, ensure a transparent and coherent equity story so the financial community understands the company's vision, mission, strategies, and business and financial conditions. These must be communicated in a uniformed manner by any of the company's employees.
- Ensure that non-financial reporting is closely aligned with the company's strategy, processes and communication.
- Ensure that as many factors as possible are included in the pricing of the company's shares and the correct formation of expectations on the company's part.
- Ensure and maintain the company's image in the capital market as a visible, accessible, reliable and professional company with a consistent flow of financial and other information.
- Create and maintain a close, open and honest relationship with current and potential investors and equity analysts to create goodwill for "a rainy day".
- Create and maintain a broad-based shareholder base, in geography, shareholder type and time horizon.
- Ensure that price-sensitive information reaches all financial markets stakeholders at the same time.
- Be proactive rather than reactive.

The IR function is continuously involved in the implementation of several internal objectives, including:

- Pass on relevant thoughts and inspiration from the financial community to the company's management and BU managers.
- Pass on important competitor information, which IR collects, to relevant parties within the company.
- Ensure that the company publishes no incorrect, misleading or inappropriate information.
- Advise colleagues on confidential and share price-sensitive matters.
- Form a qualitative part of the company's overall integrated communications policy.

Two-Way Communication

The IRO must have in-depth knowledge of how stakeholders think about the company to communicate efficiently and make the company's messaging stronger.

As the financial community is dynamic by discussing and interpreting new information, the IRO needs two-way communication with the financial community. Sometimes an IRO may be given a specific message to communicate to the market, but the IRO needs to challenge the news as the stakeholders would. This provides the IRO with better information transparency and so-called ammunition when discussing the messaging with the financial community and the company's stakeholders.

In addition to communicating key messages, the IRO must have a regular and recurring dialogue with the equity analysts and investors so that the granularity of the message is high, allowing the message to be encapsulated by the market perception and, therefore, the market price of the shares. We will later (Chapter 18) explore competitor intelligence, which is a clear area where the board of directors and senior management of a company can utilise the know-how of the IRO and the dialogue with investors into the company's best practices.

The Goal of Achieving a Fair Value

Suppose the IRO masters the main objectives of IR and has a strong positioning internally and externally. In that case, the function can reduce a company's shares risk premium and thus reduce capital costs to raise capital. This is an apparent reason why most M&A-orientated companies need to have an IRO, as they understand the benefits of financing new business ventures. However, it is not the goal to promote or inflate the share price because this increases unnecessary short-term volatility.

As the company's share price fluctuates relative to the company's internal perception of the intrinsic value, the IRO will advise senior management accordingly. We further explore communication tools and options at the IRO's disposal (Chapter 21).

A company is recommended to host a virtual webcast following all quarterly updates in order to maintain the dialogue with its stakeholders.

In connection with the webcast, the management will give more granularity to the written updates. The IRO should recommend to management in which tone the message should be communicated depending on the level of risk premium or discount of the company's share price. It is, therefore, possible for the management to put a more positive or cautious focus on specific aspects relative to the risks and opportunities of the company.

The IRO may also collaborate with the finance functions and the equity analysts to ensure that both credit and equity investors are familiar with the company's current focus and strategic direction. These ongoing relationships result in investors having a more realistic and trustworthy management and IR function. In addition, it allows the financial community to understand how the company will react if facing financial problems or if the business model allows a more proactive policy to raise capital.

It is vital to attract a diverse investor base to achieve a fair value of the company's shares. It is, therefore, a valuable practice to maintain a list of current and potential institutional shareholders and work practices to search for which PM's and equity analysts are relevant to draw their attention towards the company.

An updated "wish list" of investors and equity analysts will assist in shaping the company's IR communication and alter it to attract new investors. A clear example is that many investors now require a comprehensive ESG reporting (see Parts XII–X) before relevant PMs can conduct due diligence and investment analysis of the company.

Lastly, it is IRO who delivers the "building blocks" and financial targets to allow the investment to communicate to understand the company's direction and, if done correctly, reduce activist investors and high share volatility relative to the earnings development of the company.

CHAPTER 18

Deciding on IR Ambitions and Its Success Factors

The Basics of IR

Initially, the basics of IR should be in place, i.e. clarification of ambition level, internal organisation and division of responsibilities, the extent of senior management's involvement, handling of corporate governance-related matters, IR policies and the interaction between IR and the company's other internal and external communication. Before a company can start working with IR, some clear conditions should be in place. Once this foundation is established, the company must plan the IR activities for the next one to two years. After this, the company is ready to execute—and hopefully excel (Fig. 18.1).

The Ambition Level of IR

One of the first things a company needs to decide is the level of ambition within IR. The purpose of IR is often included in a company's IR policy and reflects a company's level of IR effort, which should be published on the company's website. In our experience, it is a recurring question how much IR effort is necessary and recommendable for a listed company. The answer is that the final determination of the IR effort depends, among other things, on:

- The company's industry—some industries are more closely followed by investors and equity analysts than other industries. The greater the focus on the industry and the company on the financial markets, the greater the need for IR efforts.
- How do the company's domestic and foreign peer companies (wholly or partly comparable companies that are direct or indirect competitors) handle their IR activities?

P. Lykkesfeldt and L. L. Kjaergaard, *Investor Relations and ESG Reporting in a Regulatory Perspective*, https://doi.org/10.1007/978-3-031-05800-4_18

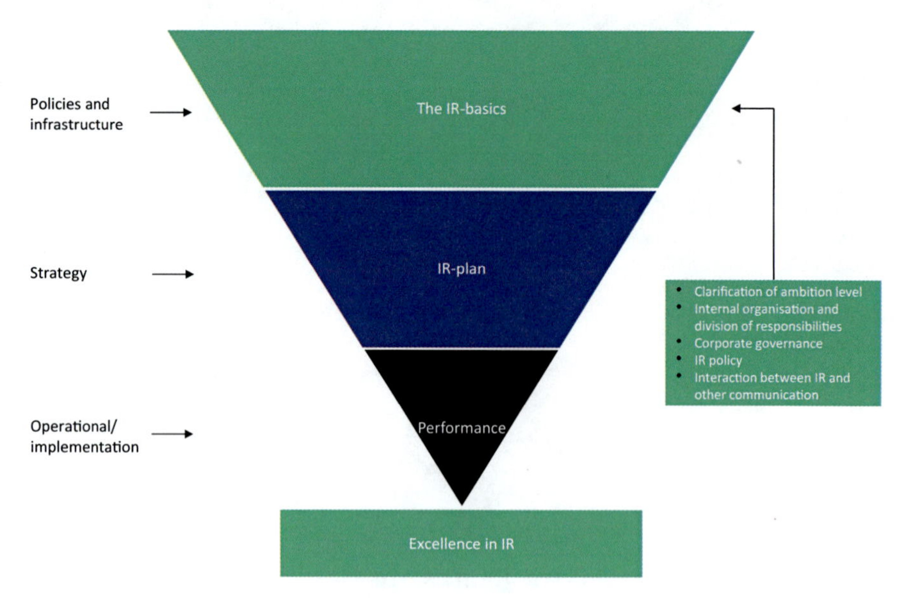

Fig. 18.1 Building blocks of best-practice IR

- What is the company's level of ambition in IR?
- Whether there is a separate IR function in the company, in the opposite case, the IR activities are typically handled by the company's CEO or CFO (typically relevant for SMEs).
- How large a budget is a company willing to allocate to the IR area?

IR is a strategic leadership discipline and is rather an art—than a science. This is one of the reasons why it is not always possible to give a clear guideline for how a given situation should be tackled, but rather an indication of which factors should be included in the company's overall considerations.

Some companies occasionally ask whether there is a need for IR at all. This is especially true if the company has delivered good financial results for several years, expects to do so in the future and has limited reliance on raising capital to fund expansion opportunities. The IR department acts as a service function, but fundamentally, IR is a strategic tool for the company in line with, e.g. HR and communication to the company's other stakeholders. Therefore, it is up to the individual company to prioritise how central IR should be in the organisation.

Prioritising the Level of IR Ambition

Should the company be best in class, on a par with the average of the competitors in the company's sector in question, or simply have an IR function of name, but without much benefit? As a starting point, the efforts in the IR area and the pricing of the company's shares go hand in hand.

The typical picture today is that with a few exceptions, all companies in the IR field are upgrading and that the quality level of the average of the IR function is constantly increasing. In connection with the prioritisation of the IR effort, and this will in most cases mean both about time and money, this is both an initial effort and an ongoing effort (Fig. 18.2).

Some larger companies can attract investor interest, even though the companies are less active on the IR side. These companies have opted for a subdued IR policy but can still gain interest from investors due to their significant weight (share) in the indices that they fall in. This is legitimate behaviour, as investors know what they are getting into when buying the shares of these companies. It can even be said that these companies contain an upside, should they later change policy in the IR area. However, the behaviour of these companies in the IR field is not advisable in the long run. Therefore, if a company wants an IR function, the management, and the IRO, must decide:

- Who will oversee the daily IR work?
- How should the IR function be staffed (number of people and competencies)?
- Total budget, including travel activity/budget.
- The quality of the IR tools (Chapter 21).

It's just a matter of the amount of resources management wants to allocate. In several SMEs, the IR function is typically handled by the company's CFO, alternatively CEO, which in many contexts is also sufficient. In any case, it is essential to prioritise the IR effort correctly so that the company optimises the return on the IR return concerning the resources used.

Fig. 18.2 Prioritising the level of IR ambition

The Requirements for the IR Officer

The background of the IRO should ideally be broad, as the role of IR employee requires competencies in the following areas:

- Knowledge of the financial markets.
- Communication.
- Finance.
- Strategy.
- Business insight.
- Scientific insight (depending on industry).
- Proactive, internally as well as externally.
- Can perform at all levels of the organisation.
- Political clout.
- Dare to challenge views internally (often management), constructively and externally (investors and equity analysts).
- A feel for sound diplomacy.

Knowing the company, industry and internal organisation is advantageous. Therefore, candidates for junior IRs position are often recruited internally within the company, e.g. younger employees from the finance area, e.g. controllers. However, a senior IRO is often hired externally, with a career as an equity analyst in the same sector, or as an IRO in peer company, i.e. with some existing knowledge of the company.

If the company's desire is a candidate who can essentially stand on its own two feet from day one, external recruitment is usually the most obvious. Ideally, a candidate may be recruited within the same industry, as industry knowledge takes time to build. Often, equity analyst who covers the company are also recruited to the company as IROs.

In any case, the IRO should have some apparent leadership abilities and be respected in the organisation and financial community. The typical institutional investor and equity analyst is highly specialised and has an excellent insight into the industry in question and the competitive situation. These external stakeholders rightly demand excellent and professional service. If these experts encounter incompetence in the IR function, it will immediately backfire, and it may in future be difficult for the company to arrange investor and analyst meetings without a member of the top management attending. In addition, there may be the challenge of dealing with IR-related issues daily.

A Regular day of an IR

It is essential to meet as many relevant investors and equity analysts as possible face to face in the form of both group and especially one-on-one meetings. In this phase, the management and the IRO build personal relationships. Webcasts are also a valuable and effective tool for attracting foreign investors and equity analysts. At the same time, a Capital Market Day (CMD) must be considered a relatively lower priority. Identification of new potential shareholders is, of course important, but not particularly time-consuming.

An IRO needs to have an ongoingly proactive (instead of reactive) approach to the financial community. The IRO should be a recurring point with the board of directors, and they should receive every month the best research reports, and an IR report, of the ongoing IR activities and price/volumes, what have been the most important meeting with investors and what were the Q&A.

Along with target price and recommendation from the equity analysts, the CFO (or the IRO) should present the report quickly so that the board of directors can ask questions and may be familiar with how the market views the company. The board of directors should wish to have regular updates on the market perception of the company.

The daily handling of operational IR tasks includes:

- Answering phone calls and e-mails from investors and equity analysts.
- Holding one-on-one meetings with investors and equity analysts.
- Maintenance of the company's IR website.
- Preparation of the company's IR plan, investor presentations, annual reports and quarterly accounts.
- Preparation of stock exchanges.
- Participation in the solution of any communication-related critical tasks and crises.

Finally, IR is typically (or legal counsel) wholly or partly responsible for ensuring that the stock exchange's disclosure obligations are complied with. These are both the primary IR tools, e.g. the company's website and annual report, but even more so the long excellent daily move of servicing financial markets stakeholders that results in excellence in the IR area. This includes a punctual response to inquiries from investors and equity analysts and the differentiation from other IR functions in the form of service to the IR function's stakeholders, investors and equity analysts.

An example: As for the subsequent annual accounts, a company must change accounting principles or structure in the company's segment information, and these changes are not considered price sensitive. What do investors and equity analysts prefer:

- They get a surprise in connection with the actual financial reporting. In contrast, a general rule is that they are swamped with interpreting and concluding the usual accounting information and suddenly must rework their models and spreadsheets? **OR**
- A week before they have received the historical data in the new format, do they have plenty of time to adjust their spreadsheets to the new layout?

The answer should probably be quite clear and will, in the latter case, lead to goodwill on the part of the users. First, however, it must be assessed in each case whether, for price-sensitive reasons, it is practically possible to provide the financial markets with this helping hand.

Competitor Intelligence

Competitor intelligence is a function within the company's strategic functions. However, IR has access to sector and specialised knowledge and a treasure of knowledge that can be passed on. A clear missed opportunity if a system is not in place if the information is not passed on within the organisation.

The IRO typically receives a large amount of analysis and reports from equity analysts, not just about his own company but also about competitors and the industry. In addition, these reports can be precious to colleagues elsewhere in the organisation from a competitor intelligence point of view. In this context, it is recommended that internal procedures be established to ensure all internal stakeholders benefit from these analyses (Fig. 18.3).

Fig. 18.3 IR's role in competitor intelligence

IR is usually the recipient of the internal information, either daily ad hoc, answering or answering questions from investors and equity analysts. In addition, their role is also to collect company information in connection with the preparation of investor presentations, annual report or annual and quarterly reports. Therefore, it can be a clear strategic benefit for a company to have an IRO with a robust relationship with the financial community and internal stakeholders.

In addition to creating goodwill in the organisation, as the connections can lead to value-adding business opportunities, the IRO provides important advice to senior management or the company's strategic direction. Before commencing on analyst and shareholder targeting, a company should fundamentally wish to build a good investor base supporting long-term objectives.

Targeting of Equity Analysts

We estimate that best-practice IR versus unfocused IR may lead to a significantly reduced risk premium. A key reason is that best-practice IR leads to better investment coverage from equity analysts and, therefore, greater attention and information disclosure towards investors. For example, a typical equity analyst discusses company updates with 15–25 tier 1 investors. In addition, the equity analyst also keeps the brokers informed, who can together discuss the case with 50–100 institutional investors on any given day. This multiplicator effect is a clear benefit from onboarding one equity analyst who can communicate the critical information changes for the company.

In addition to discussing the information with investors, they also provide regular commentary to the media, which likewise has a high multiplicator effect, especially towards retail investors. These multiplicator effects are only possible if the company is covered by an equity analyst and is more profound the more equity analysts cover the share.

Therefore, one of the main IR objectives should be an annual target to increase the number of equity analysts covering the share. This should also include a constantly updated "wish list" of equity analysts to follow the company. This list can consist of equity analysts who follow similar peer companies, work with them and keep them engaged to find it easier to follow the company. The IRO should also proactively contact new equity analysts between its trading and quarterly updates, where the equity analysts have more time, to ensure they are well informed.

As brokers are remunerated from the volume and price of trading shares, the larger the company, the more the traded volumes. It is, therefore, more accessible for large companies to attract equity analyst attention because it is easier for the investment bank to make the cost–benefit analysis. On the other hand, it is more difficult for SMEs to attract independent research coverage, so they may consider a commissioned research model from a quality-orientated investment bank.

Commissioned research is by no means a sunk cost. It can also complement independent research, as they both assist in the formation of information surrounding the company and thus a lower risk premium. Following MiFID II of 2018 (Part III), the field of commissioned research has gained popularity among investment banks and institutional investors—as well as retail investors.

Shareholder Targeting

In addition to a "wish list" of which desired equity analysts the company could be covered by, the IRO should also have a long- and short-term strategic plan on investor targeting. This includes uncovering the most relevant themes that the company is exposed to, and which investor mandates operate within the space. Most investors have a long screening list of potential investments but only have very few (15–30) companies in their portfolios. Therefore, a proactive IRO in collaboration with the equity analysts could open doors to expand the shareholder base.

Most investors focus either on high-growth companies or on companies they believe are undervalued relative to their actual values or long-term potential. The IRO should therefore make an extra effort to get these investors to acquire a long-term shareholding in the company. One should generally seek to build a broad (shareholder types), stable and long-term circle of owners of satisfied shareholders. This also includes retail investors, who can play a significant role in the liquidity of the company's shares.

The advantage of a satisfied shareholder base is that it increases the probability that it will support the company on the day it has to raise capital at short notice by issuing new shares for an acquisition, or on the day someone tries to take over the company via a bid at a price which is deemed to be below the fair value for the shares. If shareholders trust management, they are more likely to support management's recommendations. Is there, e.g. in the case of a takeover bid for the company's shares, they will be more likely to allow any doubt to benefit management, which may ultimately lead to a wholly or partly group of shareholders rejecting a given takeover attempt.

When the company engages in investor events such as regular investor meetings, roadshows and other IR activities, typically hosted by an investment bank, the IRO should request a participants list. Keeping track of the participant list is helpful for shareholder targeting and a KPI (key performance indicator) in board meetings. In addition to registering who participated in the discussions, the IR should note questions, key concerns, and ideas during such meetings.

Complementary to regular meetings, the same guidelines should be kept regarding strategic CMD, webcasts, one-to-one meetings, group meetings, etc. We will explore these dynamics in Chapter 21.

CHAPTER 19

IR Within the Organisation

Reporting to the Senior Management and Board of Directors

Good corporate governance involves the principal agency theory, in that the senior management ("the agent") and board of directors ("the principal") have asymmetric information; both functions act to maximise their utility, and their goals are not always aligned. The agents tend to be more opportunistic. Therefore, the board of directors also has clear benefits of utilising the IR function. However, a short-term CEO may be less transparent towards the board of directors, as they are his employers.

Having a stable connection with the IRO allows better strategic and shareholder transparency, which satisfies current shareholders to a greater extent. However, a sense of realism is vital towards the senior management and board of directors with clear communication surrounding the market perceptions.

The board of directors has a clear responsibility to advise senior management on strategic direction, significant corporate changes (e.g. M&A), dividends and matters associated with raising and issuing new capital. If an IRO can remove just 1 percentage point of risk premium in a company's share price valued at 1 billion euro, this equals a rate of return on IR of 10 million euro. Therefore, the investors, thereby the board of directors, should have a clear interest in the strategic decisions and KPIs for the IRO.

P. Lykkesfeldt and L. L. Kjaergaard, *Investor Relations and ESG Reporting in a Regulatory Perspective*, https://doi.org/10.1007/978-3-031-05800-4_19

Internal Organisation and Division of Responsibilities

As the IR function combines the knowledge of various roles within a company and relays it to the financial community, it has a limited formal organisational process and, therefore, little formal corporate responsibility.

However, informally, IR has a wide range of connections and influence within the organisation due to the strategic and operational importance of the function and the support of the senior management. As a result, a typical IRO career path is straight either from a financial function into IR, or from the position of a former equity analyst position. Some IROs continue their careers within IR in the longer term, while others after perhaps 3–5 years as head of IR continue their careers in the same company, e.g. line management or in a higher position in finance—in both cases perhaps eventually reaching a senior management position some years down the road.

Besides the dedicated IRO, senior management also has a clear role when practising IR. In the best interest of all stakeholders, that senior management dedicates most of the time to the business itself. However, statements of consideration from the CEO/CFO are always more respected in the financial community. The equity analysts and investors may even save their more in-depth value add questions to the CEO/CFO so that the IR has not prepared answers for them beforehand. Prominent institutional investors also require regular updates from the management and not only from the IR. It is the IRO's responsibility to assist prioritising the time of senior management and allocate time resources appropriately to IR activities.

IR's Reporting Line

IR is an area that should have excellent attention and direct reference to senior management and is a support function for the CEO/CFO. The specific reference point for the IRO is the company's CFO, alternatively CEO. It is also often seen that integrating the IR function in the communications department—or vice versa—strengthens its overall communication competence especially in companies where the head of all communications is part of or part of, if not the senior management, then its upper management team. Regardless of the reference conditions, the IRO must maintain a very close collaboration with the company's CFO.

- The benefit of reporting directly to the CEO is maximal credibility with investors, along with the chairperson, the most influential person in the company. In addition, it allows the IRO to better keep the CEO up to date about market sentiment when the CEO is communicating with investors. On the other hand, the IRO's daily work has a higher synergy with finance and communication and is, therefore, less combability to the daily work of the CEO.

- The benefit of reporting directly to the CFO is higher synergies with the daily work and a higher ability to keep close to the finance area.
- As external communication is central for IR, the benefit of reporting to the communication department is a more up-to-date view of how the company is communicating to its external stakeholders at a given time. In addition, it allows synergies of the overall external communication of a company. On the negative side, an IRO not reporting directly to the CEO/CFO will often have slightly less credibility in the financial markets, all things being equal.

The organisational reporting point of the IRO is thus something that must be decided on a case-by-case basis and where one should think carefully about. There are several obligations associated with being a listed company. The company is also obliged to prepare internal rules:

- To ensure compliance with the disclosure obligations.
- To ensure that inside information is not passed on unjustifiably.
- For trading in the company's shares.
- For insiders trading in the company's shares.
- Silent periods (i.e. a temporary break in the dialogue with the financial markets before publication of financial statements).

Board of Directors' and management's Role and Involvement in IR

In the following, we assume that the company has a so-called two-tier governance structure, i.e. a separate board of directors and senior management, and that the CEO is not also the chairperson of the company, which in a wide range of countries is neither allowed by law nor by investors considered best governance practice.

Generally, we recommend that senior management runs the company's IR activities. In our view, this provides management, and the IRO, with the relevant credibility among investors and equity analyst. However, it is common that the largest investors and shareholders have annual meetings with the chairperson. This is both warranted and relevant as ultimately, it is the board of directors setting the direction of the company, not the CEO. Further, if the larger investors and shareholders for good reasons are unsatisfied with the CEO or other members of corporate management, then naturally the larger investors and shareholders may reach out directly to the chairperson requesting a meeting.

We recommend, however, that the IRO offers a structured monthly reporting to both the board of directors and senior management, which includes, e.g. the following:

- Share price development, commented
- Volume development
- Equity analysts' consensus estimates
- Overview of equity analysts' estimates, recommendations and target prices, including changes from previous month
- Summary of published equity research
- New analysts and targets
- Key investment themes from equity analysts—positives and negatives
- Key meetings with shareholders (potential and existing)
- Top 25 shareholder movements
- Overview of IR meetings
- Short interest (% of outstanding shares), including development among potential activist investors or potential bidders.

In the event that the company has its own IRO, then it is a balance how much time, e.g. the CEO and CFO have to engage with investors. In conclusion, senior management also need time to run the company. Here, executive simply have to prioritise as investor and equity analyst often prefer senior management interaction rather than IRO interaction (unless they need a detailed and quantitative run-down).

Hence, although the IR activities on an isolated basis may benefit from endless senior management involvement, then at some stage the overall company starts to suffer if senior management, e.g. the CEO and CFO, do not have the time to run the company. We seek to illustrate this theme in the below illustration (Fig. 19.1).

Collaboration with Internal Line Managers

In our experience, most business professionals love to discuss their work. A key strength of IR within the organisation is that it has no business responsibility yet has a solid basic knowledge of the company's strategic direction and current financials. Therefore, an IRO will create a strong network within the organisation. The IRO can also relay information about financial markets' view on the company and support line managers without being their superior or subordinate.

This relationship is only really constructive when based on a mutually respectful give-and-take relationship, e.g. providing some competitor intelligence (Chapter 18). However, if the IRO is too arrogant or disrespectful, the line managers may only provide minimum details. Therefore, it is recommended that the IRO creates a positive internal network and upholds respect, allowing better information to flow in each direction. This also allows receiving specific (non-material) anecdotes from the line managers, which the financial community appreciates.

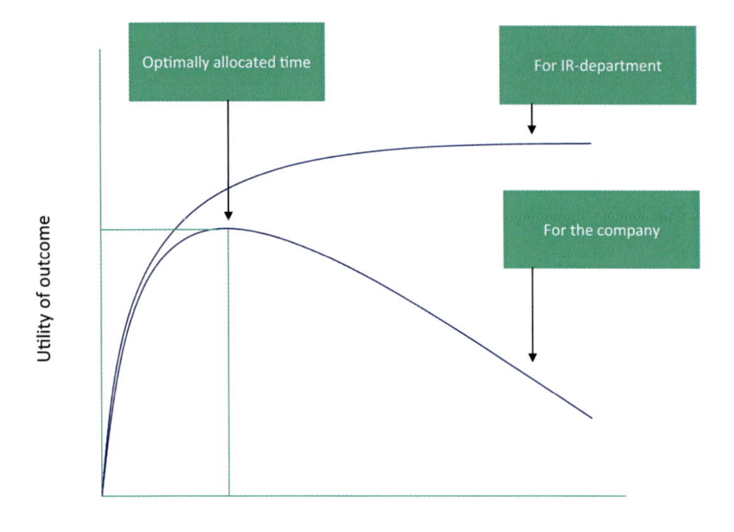

Fig. 19.1 Senior management's time/utility of outcome on IR

Collaboration with the Communication Function

The company's internal and external communication must be carefully coordinated to ensure consistent communication to the company's stakeholders, regardless of whether the company reports to the outside world in a more traditional way or according to best ESG practice.

The advantage of close cooperation also applies to handling selected financially and business-oriented media. Navigating the disclosure of information to the financial community is where the IR function has its primary competencies. Although this target group of stakeholders typically falls under the communications manager, the IRO may occasionally take on the role of spokesman in these contexts due to the financial communities' focus on these media (Fig. 19.2).

At the time we write this book, most corporations are scaling up on their ESG functions. IR and ESG should be either formally or informally integrated as non-financial reporting gains importance within the financial community. It is the responsibility of IR to coordinate the quarterly results report and thus acquire which ESG data and granularity are relevant to communicate. As the financial community is primarily concerned about changes in the information, the IRO needs to have a clear overview of its ESG goals, targets and progress. Publishing wrong, misleading or confusing ESG data will have negative sentiment consequences just as it is the case with reported financial numbers.

The IRO needs to create and maintain control mechanisms to receive and check incoming ESG KPIs quickly. In addition, when external stakeholders

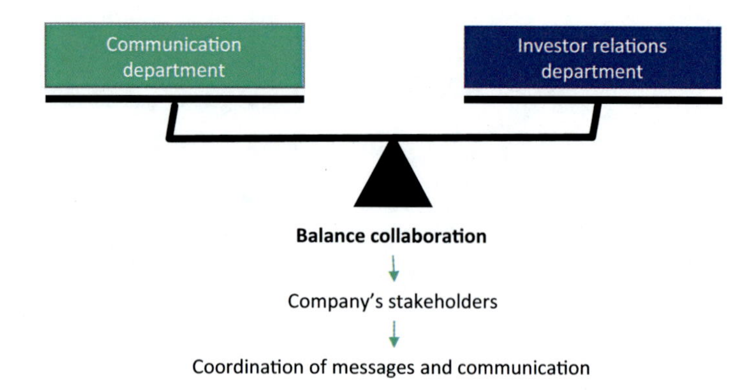

Fig. 19.2 Collaboration between the IR and communication departments

request specific data, it is essential to promptly deliver the data to the market if the IRO finds it relevant. Proactive and positive collaboration between IR and the ESG and communication functions will be seen among ESG investors as a clear positive signal.

However, it seems that there are currently more positive examples of proactive ESG approaches in the financial markets than reluctant integration of these standards.

In terms of both the practice of IR and ESG reporting, Denmark has been on the forefront over the past two decades. On this background, we wish to highlight three Danish examples of proactive ESG, which we believe well symbolises a very proactive ESG strategy, which many IR functions can learn from.

Examples of Proactive ESG Approaches

Novo Nordisk

Novo Nordisk is the largest company in the Nordics (by market capitalisation) and has kept sustainability high on its agenda for several decades and was the first to publish a sustainability report at that time referred to as an environmental report. In 1994, the global insulin producer and diabetes care company published an environmental report and, in 1997, introduced "the triple bottom line", internally championed by Lise Kingo. The term was initially coined by business writer John Elkington in 1994.[1] The triple bottom line is aligned with financial, environmental and social responsibility.

As Novo Nordisk (which at that time also contained the bio-industrial activities today known as Novozymes, which was spun-off and separately listed in year 2000) was a pioneer for non-financial reporting, the report itself was a key global driver of sustainability. The information originated from the bio-industrial segment that focused on genetic technology. Given NGO pressure,

Novo Nordisk decided to embrace them, provide transparency and host seminars for the NGOs to discuss their issues. Therefore, the report itself was a significant milestone but was complemented with stakeholder and employee integration. As a result, Novo Nordisk could balance short-term economic motives with long-term societal interest throughout its business practices.

The spin-off of Novo Nordisk, Novozymes, was the first company to produce an integrated report in 2002 with non-financial and financial reporting, a standard later adopted by Novo Nordisk in 2004. Denmark made this obligatory in 2012 under the NFRD (now CSRD) outlined in 2014.[2]

Rockwool International

The global producer of stone-wool insulation, Rockwool International,[3] was the first company to initiate quarterly ESG conference calls in 2019. In connection with each quarterly statement, the company found that more questions from investors and equity analysts concerning ESG arose on its webcasts. As a result, the company introduced a separate call to be an in-depth perspective on a particular theme and allowed an elevated focus on ESG.

At the time of writing this book, Rockwool International has now hosted ESG calls for ten consecutive quarters. Insulating buildings and energy efficiency are crucial to reducing CO_2 emissions and, therefore, a very obvious ESG topic. However, we believe Rockwool International's specific approach is impossible for all companies to copy. The themes on Rockwool International's webcasts include, e.g. "climate changes and energy efficiency", "circularity", "social impact and governance", "ESG in the value chain" and others.

AP Moller Maersk

Maersk is the second largest company in the Nordics (by revenue) and the most extensive container shopping and vessel operator globally. As shipping is inherently an industry with high CO_2 and NOx emissions consumption, Maersk has over the years engaged in dialogues with the financial community about the industry's transformation. This includes producing the first methanol fuelled container ship (and ordering eight of them in 2021) and investing in PtX (power-to-X) technology, transforming green energy into hydrogen-based fuels. The company has come a long way in recent years in terms of ESG reporting and has initiated a proactive approach despite Maersk's sector having a high CO_2 footprint.

In 2018, Maersk committed to the Paris climate accord to become CO_2 neutral in 2050 and therefore viewed the green ambitions as a growth opportunity rather than a negative industry disruption. Yet, in the beginning of 2022, the CEO announced that it aims to reach carbon neutrality by 2040 and set a 2030 milestone of reaching the science-based targets. Following the announcement, the company also announced to host its first ESG investor day on 3 March 2022.

Today's new methanol fuel container vessels cost 10–12% more than standard container vessels. However, large companies, including Amazon, IKEA and Unilever, have already announced that they wish to use green vessels by 2040 and pay up to 10% higher than the alternative. Therefore, it is widely expected that Maersk will align its goals with these statements.[4]

References and Sources

1. Elkington, J. (2018). 25 years ago, I coined the phrase 'triple bottom line'. Here why it's time to rethink it. *Harvard Business Review.*
2. Muga, H., & Thomas, K. (2013). *Cases on the diffusion and adoption of sustainable development practices*. IGI Global.
3. Website: www.rockwool.com/group/about-us/investors (Downloaded November 2021).
4. Finans.dk, 3 November 2021.

CHAPTER 20

IR's Responsibilities of Implementing Policies and Planning Ahead

The Annual Reporting Wheel

According to the IR-service, Q4 Inc., on average 20% of all share price movements are attributable not to announced financial information but to what is said in terms of corporate communications. The IRO has a clear role in planning senior management's calendar to best prioritise its time talking to the financial community. Besides the ad hoc meetings and roadshows, it is common to meeting investors and equity analysts following the announcement of quarterly reports. The main objectives are to communicate the equity story, manage shareholder expectations and to constantly seek to increase coverage from equity analysts.

Most companies have a 4–6-week silent period prior to the announcement of quarterly results. Most tier 1 institutional investors and equity analysts appreciate a "pre-close" call with the IRO or management to match market expectations with the company's current thoughts. Subsequently, the IRO is recommended to arrange roadshows, typically in collaboration with an investment bank, every quarter in addition to hosting a webcast. We set out below an overview of the typical annual IR activity wheel (Fig. 20.1).

Aide Memoire

The IRO will publish, and make available, a public aide-memoire before going into silence. The aide-memoire is published four times a year and includes the most relevant expectation management material the IRO wishes to highlight. The main topics typically consist of:

P. Lykkesfeldt and L. L. Kjaergaard, *Investor Relations and ESG Reporting in a Regulatory Perspective*, https://doi.org/10.1007/978-3-031-05800-4_20

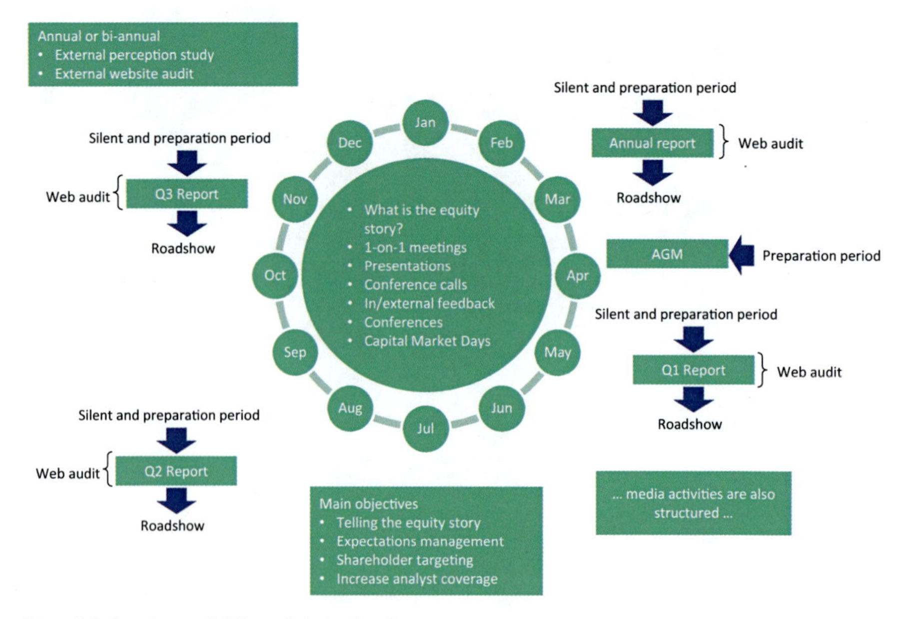

Fig. 20.1 Annual IR activity wheel

- Financial highlights of the previous report and focus on the critical takeaways on revenue and costs. It is also recommended to include information on the liquidity and debt situation and any announcement of new products or contracts.
- Financial guidance for the full-year and a commentary on the building blocks to achieve the high-end or the low-end of a possible earlier announced financial guidance range.
- Extraordinary items to consider from the previous year or items for the upcoming report. In general, it is recommended to comment on selected events the previous year which influence the base of comparison, as equity analysts and investors tend to focus on the development of the current year. Hence, if a company last year was faced a specific tailwind or headwind, it is wise to highlight it.
- Highlighted quotes from management on the previous investor calls.
- Other topics include dynamics in the market to be aware of, phasing of specific contracts, production shutdowns, weather, etc.—all depending on the specific industry in question.

IR Plan

A company's internal IR plan, which typically extends one to two years, expresses the company's planned activities in the IR area. The IR plan should also include several operational objectives, for example:

- Identification and attraction of new targeted shareholders.
- An increase in the share of private shareholders.
- Increase in the number of equity analysts covering the company's shares.
- The number of meetings held with investors and equity analysts.
- In which countries the company will conduct roadshows.
- Further development and improvement of IR tools, including an annual report, website, standard and theme-based investor presentations.
- An improved rating in a recognised independent IR study regarding the quality and performance of the company's IR function and activities.

All operational objectives should be quantified. The company will often learn that the financial markets, especially equity analysts, has wishes in the company's IR area that go beyond what it plans or finds appropriate. In such cases, the company should not blindly comply with the wishes of the financial markets but seek to assess relevant pros and cons. The choice and scope of activities is necessarily a function of:

- The company's characteristics and industry.
- Ambitions in the IR area.
- Budget.
- Prioritisation of the IR effort concerning what is assessed to give the best return on the resources used.

IR Policy

A vital area of importance for an IRO is to have a red thread throughout the company's communication towards its shareholders and other stakeholders. This allows external interest groups to follow the development of the company in a transparent manner. A company's external IR policy expresses the company's overall intentions and goals within the IR area. A company should have an established policy in the IR area. IR policies are individual, but one should as a rule draw inspiration from what other companies are doing. It is not our intention to develop a definitive proposal for an IR policy but rather to encourage companies to establish or update an IR policy and to conduct some field research.

As mentioned in Chapter 3, an excellent place to start is on major companies or their peers' websites—the IR section. Other examples of IR policies can be found on the NIRI (National Investor Relations Institute). However, access requires membership. Most of the objectives we mention in this part are often included in a company's IR policy, often freely available on the company website.

Financial Guidance and Operational Objectives

Financial guidance and operational objectives should be clear, realistic and relevant. The guidance area requires careful consideration, and a company should tread carefully. Publishing forward-looking operational objectives and guidance is the best method to manage expectations for a company. An unexpected profit warning, i.e. a downward adjustment of a previously expected result, is not soon forgotten by the financial markets. So instead, opt out a little cautiously, but of course, still within the scope of realism.

The objectives can be short-term guidance for a given year or long-term goals for the strategic direction. Providing solid guidance and objectives portrays professionalism and visibility into the company. The granularity depends on the company and sector. For example, a newly listed biotech company has limited short-term finances to guide in the short term but can perhaps highlight long-term objectives and aspirations. It is recommended financial targets published by a company may fall within the areas, however, still dependant on industry in question:

- Growth in revenue (including considerations on volume, price, M&A and foreign exchange impact).
- Profit margin, EBIT (operating profit) or EBITDA (operating profit before taxes, depreciation and amortisation).
- CAPEX and cash flow.
- ROCE (return on capital employed) or other variants of this profitability target.
- Return on equity (however of diminishing importance).
- Matrix sales targets include customer churn, recurring revenue, cost of acquiring a customer, customer lifetime value, net promoter score, etc.
- Non-financial reporting targets in the ESG area.

Several other examples of financial targets and sub-targets could be mentioned, and it is also often seen that a year is linked to when these targets are expected to be achieved. However, before the goals are announced, it is essential that the board of directors and the senior management thoroughly analyse the norm in the industry, what the peer companies do and what is realistic for the company in question. These goals, which may have an element of the stretch target, which means one or more ambitious goals, also stimulate the board of directors and senior management to realise the potential that the company contains (Fig. 20.2).

The danger with financial objectives and the desire not to disappoint the financial markets are that short-term goals are sometimes sought to be achieved at the expense of the company's long-term goals, profitability and competitiveness. We have experienced that companies after years of good results suddenly underperform. This may lead them to underinvest in their

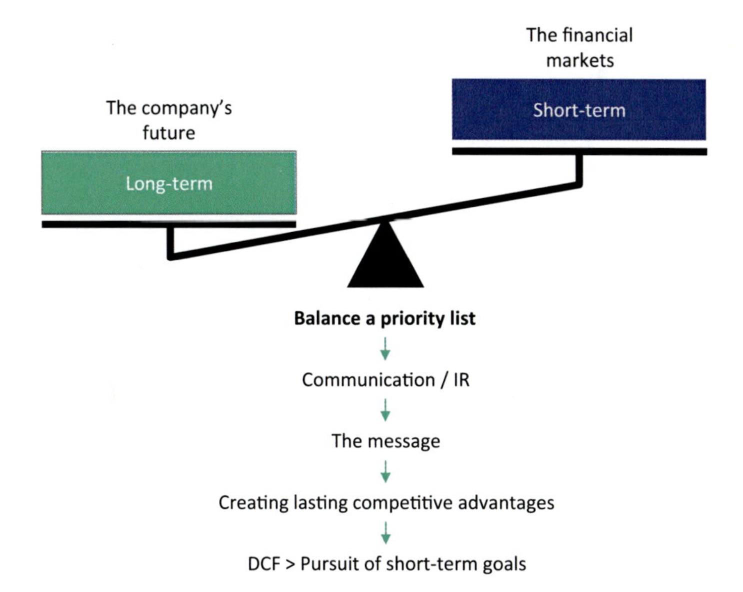

Fig. 20.2 Long-term targets and short-term guidance

long-term plans resulting in a strategic backlog compared to competitors. This is an unfavourable situation for all parties.

There are countless examples of companies destroying long-term value by complying with the financial markets' short-term demands and desires. Therefore, one of the senior management's (and the board of directors') main tasks is to ensure understanding of the need for ongoing investments in long-term business development in the financial markets. At the same time, management must ensure that these investments result in lasting competitive advantages. The value of the long haul should thus be apparent to investors and equity analysts concerning the company in question. This is a critical debate, where the board of directors, senior management and the IR must be involved.

In connection with the company's guidance (i.e. the company's publication of a given expectation for a given area—often a financial ratio—related to a given period or for the current financial year) to the financial markets, it is recommended to seek to be dynamic and nuanced. Consideration must, of course, be taken to comply with the financial markets' desire for a level of detail. Still, it is nevertheless to do oneself a disservice if one is tempted to give too detailed a guidance or press the lemon too hard not to disappoint the financial markets.

The rules in this area are quite clear about the company's ongoing management of financial markets expectations for both the annual and quarterly results. All communication must take place via announcements to the stock exchange, or legitimate electronic newswires. As established in Part I, selective disclosure is illegal. Despite these circumstances, an IRO can succeed in practising expectations management in a way that takes place within current rules and has the positive effect that unnecessary fluctuations in the company's shares are avoided. The IRO can do so in connection with the quarterly results referring to a market consensus as a way of indicating to equity analysts that their estimates may be too high or too low.

In addition, the IRO may point to those of the equity analyst's or investor's assumptions that may deviate from the general market consensus. Finally, factual errors in the assumptions of equity analysts that can be demonstrated concerning publicly available material should be pointed out. For example, the existence of a market consensus estimate, an average of different equity analysts' estimates, naturally presupposes that more equity analysts cover the company's shares.

Finally, safe harbour provisions (a legal disclaimer regarding statements relating to the future) are a must in connection with anything related to a possible prediction of future developments, financial or non-financial. These disclaimers should be considered so that the outcome of the accounting results does not deviate highly from what the management had previously anticipated.

Keeping Up to Date on Disclosure Requirements

Disclosure requirement rules typically appear the relevant "disclosure obligations for issuers of shares" of the stock exchange that the company is listed on. Of course, the board of directors and senior management are ultimately responsible for compliance with these rules. Still, daily, it should be clarified whether the operational responsibility lies under the umbrellas of the CEO, the CFO, the IRO or the legal department. Finally, the IRO of the company should take the initiative to establish and maintain the IR function's systems, procedures and infrastructure, including systems and databases, to:

- Register and follow the company's shareholder base (an external independent company typically maintains the overall shareholder register).
- Record all conversations and meetings with investors and equity analysts (this provides a good system and security to be later able to document a course of events in connection with, for example, rumours in the financial markets). In addition, such a database serves as documentation of the IR department's activity level.
- Register and internally distribute received company and industry research reports from equity analysts.
- Register the equity analysts' estimates for the company's earnings, etc., so that the company has a clear overview of the consensus estimates that exist at any given time (the average of the equity analysts' estimates).
- Identify attractive potential shareholders, also referred to as investor targeting.

For the IRO's work and credibility with investors and equity analysts, the IRO must be well updated about all the company's matters. Not, of course, that everything must be disclosed externally, but it is essential for the IRO's sense of security in external contact. Therefore, to ensure a satisfactory ongoing update of the IRO on company's matters, it is recommended that the IRO regularly receives copies of board of directors and senior management minutes. The IRO is also recommended to participate, if not in all, then in selected board and senior management meetings, or agenda points. This should especially be for discussions that relate to finance, investments and accounting matters.

Finally, it is recommended that the IRO gives the board of directors a brief update on the financial markets' view of the company before each board meeting, including information on the level of market consensus estimates. This ensures that the board of directors is constantly well informed about this essential matter.

CHAPTER 21

IR Tools to Engage a Company's Stakeholders

A Best-Practice IR Engagement with Stakeholders

The IRO should have a clearly defined role in an organisation with a specific purpose, ambition level and responsibilities—and one who leaves an overall good impression with the financial community.

Therefore, an IRO should be engaging and should have daily availability to the financial stakeholders, as well as answer basic questions satisfactorily. All comments and questions from investors and equity analysts should naturally be taken seriously, and answers to pending enquiries should be returned quickly. This chapter discusses the practical engagement and IT tools available for the IR function to communicate with its stakeholders.

Internal Q&A Material

One of the essential internal IR tools is the company's Q&A material. It is an internal, strictly confidential reference material consisting of expected/hypothetical questions from investors, equity analysts, journalists, etc., with associated prefabricated answers. The purpose of the Q&A material is to align the company's answers to its external stakeholders. However, it is also to brief the board of directors, senior management, BU heads, IR, the communication area and other internal spokespersons updated regarding the more in-depth internal answers/facts (the difference should be marked).

The internal Q&A material must ensure that everyone who speaks publicly on behalf of the company delivers the same consistent message and follows the external statements and "party line" that the company finds correct to have. The Q&A material also ensures a constant update of IR at a more detailed level.

P. Lykkesfeldt and L. L. Kjaergaard, *Investor Relations and ESG Reporting in a Regulatory Perspective*, https://doi.org/10.1007/978-3-031-05800-4_21

IR should be responsible for the preparation and coordination of this material. We recommend that the Q&A document be divided into relevant themes, e.g. strategy, finances, tax, administration, etc., and describe themes such as "margin development", "competition" and "pricing trends"—depending on the company's industry. The internal Q&A material is typically updated once a quarter with input from, for example, the following areas:

- Group matters, including legal issues.
- Finance (accounting, cash flow, investments, currency, tax, etc.).
- BU 1, 2, 3 ... (with an appropriate segmentation and relevant information about the company, its competitors and the market as a whole).
- R&D (research and development) if relevant in the industry in question.

In addition, it is recommended that IR and the communications department, as a routine, prepare a set of Q&A in connection with all significant press and company announcements. The document's intent is to be on the forefront of questions from the financial community and ensures consistent communication from all internal spokespersons. In addition, having a clear Q&A document limits the IR function of collecting pending enquiries to be returned to, i.e. limiting the use of the comment "I will get back to you".

Annual Report

The annual report is a very important IR tool—and the most important source of information for equity analysts and investors. The annual report is the company's primary business card for the financial community, along with a website. Therefore, the company should spend sufficient time to ensure a good result in this area. The annual report is typically prepared in collaboration between the IR department and the communications department by guidelines laid down by the board of directors and senior management, and will possible supplementary input from external consultants, e.g. IR consultants, graphical designers and photographers.

There are many considerations to consider in preparing the company's annual report due to many internal and external stakeholders. As mentioned in the introduction (Chapter 17), this book's purpose is not to provide the recipe for the overall integrated communication policy. However, we stress the importance of always considering the appropriate stakeholders when preparing the annual report. This typically works in a collaboration between the IR department and the communications department with subsequent reporting to senior management. It is generally recommended that the planning of the forthcoming annual report begins immediately after the company reports the result for the second quarter to establish a steering group and relevant working groups. Some companies even commence with a slow start after the announcement to the first quarter results.

From an investor's point of view, the company should be careful in preparing, in addition to the accounts themselves, the following areas in the annual report:

- Letter to shareholders.
- Description of the equity story, i.e. why it is considered promising to invest in the company in question.
- Financial goals.
- Corporate governance-related matters.
- An appropriate segmentation of revenue and EBIT, respectively.
- Description of the business conditions, the competitive situation and the overall market development. The report must contain an appropriate degree of quantification.
- Research and development in the company and the industry.
- Shareholder relations.
- Non-financial reporting, including ESG goals, progress and targets.
- For non-financial reporting, the company should consider integrating ESG and financial performance in their communication.

It is advisable, under no circumstances, not to sweep any negative matters under the rug but address them appropriately in the annual report and the company's other IR tools. In this way, the company gets the opportunity to get its message across and thereby help set the agenda and premises vis-à-vis the financial community in connection with the discussion of these matters.

The trend is towards providing a more detailed description of the overall competitive situation, an illustration of the closest competitors, and relevant market shares and growth rates for the company's markets.

In addition, the financial community values consistency. This means that investors and equity analysts are very keen on following the development and see the company follow up on previously set goals and information. Similar, it will not be appreciated if the company runs a zigzag course and frequently changes financial goals and segmentation in the report, simply because the results obtained look more appropriate for the company. This behaviour will immediately be seen through and punished by investors, equity analysts and the financial media in general. In the latter respect, listed companies should be aware that in several cases, close ties have developed between selected journalists and equity analysts, and there is always an equity analyst with either a buy or sell recommendation who sees it as reasonable to have his views aired through the financial press, either with a quote or anonymously?

It can generally be said that in connection with the preparation of the annual report, investor presentations and websites are interesting to see what the competitors are doing and see what companies in other industries are doing. It is recommended to investigate the foreign development in the field, including best-in-class companies. One may very well learn from others and further develop the IR work to become the best in class. If the company has a shareholder magazine, one should also make an extra effort in its preparation.

Quarterly Announcements/Trading Statements

It is not necessary to communicate the equity story in its entirety for quarterly report but just provide a short recap. The focus should be on the news and results, and not repetitive former reports. It is therefore critical that the company offers quarterly three-month numbers. Some companies only provide six-month, nine-month or even YTD (year-to-date) numbers, reflecting performance in already announced numbers. We believe these may be included separately, but the focus should be on three-month numbers. If there is nothing significant to report on specific items on a three-month basis, then it is reasonable not to include information for the reason of filling out space.

Our most important advice for quarterly reports includes:

- First, create a good and illustrative format—let the quarterly reports be easy to read, preferably with regular and consistent graphs and tables.
- Incorporate the same segmentation that is included in the annual report.
- Be consistent and follow up on the company's equity story.
- Be consistent and follow up on previously mentioned critical issues. Address the problems that interest investors and equity analysts, and do not sweep negative anything under the rug.
- Issue, as the first option, the quarterly report before the stock exchange's opening, alternatively after the trading of the stock exchange is completed. This facilitates the evaluation of the quarterly accounts for the financial markets' stakeholders and ensures more appropriate pricing of the company's shares in the light of the new information.
- Ensure availability of management and the IR function immediately after the publication of results.

The above points help create confidence in investors and equity analysts and give the impression of professional management that is not conflict-averse or manipulative.

Other Stock Exchange Announcements

It is essential to find a balance for this frequency concerning other stock exchange announcements. Like guidance, they should be clear, realistic and relevant. It is sometimes seen that some companies publish news that is without or with minimal significance for the price formation of the company's shares. In such cases, the company's management can lose some respect in the financial markets. The financial markets would like to be kept informed, and it is good for the company to have some airtime in the market, but the news must be relevant. A company announcement should only be published if a reasonable investor finds relevant information when investing in the company—otherwise, the news should be announced in the form of a press release, not a company announcement, which are the ones, which the financial markets are focused at. With social media, the speed and amount of breaking-news content are increasing, so the IRO must be quick to determine which news are material and which are not, and decide the speed and level of granularity with which to communicate to the market.

Investor Presentations

The company's investor presentation is the most crucial educational material to onboard new (or former) investors and equity analysts, and keeps current shareholders informed. The material should include a detailed equity story, a description of megatrends and industry dynamics and the company's objectives. The equity story should be straightforward so that investors can revert to the presentations some years in the future—the company's investor presentations are typically prepared in PowerPoint, and subsequently converted into PDF format.

It is recommended that the company has organised several standard presentation templates, which can be continuously updated, in the following areas:

- Presentation for use in connection with the annual financial statements.
- Presentation for use in connection with the quarterly accounts.
- General presentation for use by investors and equity analysts who do not have in-depth knowledge of the company.
- Advanced presentation, which goes a step further, for use by investors and equity analysts who have an in-depth knowledge of the company.
- Presentations on various topics in more depth, e.g. various business areas and competitive conditions, selected geographic markets, research and development, outsourcing, industry trends, etc.

The structure of the company's overall presentation may look as follows:

- Safe harbour provisions.
- Company at a glance.
- Management.
- Summary of the company's activities and some key figures.
- Equity story/investment case.
- History and background (very brief).
- The company's markets.
- Products and services.

- Research and development.
- Summary of the competitive situation.
- Financial matters and key figures.
- Recent developments.
- Expectations for the future.
- Q&A.
- Appendix (selected background slides).

When assessing the above, it must, of course, be seen in the light of the company's size, activities, industry and ambitions in the IR area. However, we believe this is a very high "pay-back" associated with the preparation of good investor material. In addition, it provides several structuring talking points when meeting representatives from the financial markets. Finally, the more the company can set the agenda, the better the equity story.

Whether an actual manuscript should be prepared for internal use in connection with, e.g. quarterly website presentations to investors and equity analysts, is a question of temperament and training. Still, it is generally not preferable from an investor and equity analyst point of view, as the presentation becomes too stiff and formal if senior management merely reads aloud. However, the preparation of a manuscript, notwithstanding that it may not be used to hold meetings with investors and equity analysts, will always contribute to an increased focus in the presentation and must be considered a valuable discipline.

The overall company presentation can be appropriately included in the company's investor and press kit, which institutional and retail investors, equity analysts and other interested parties can request.

Furthermore, it is recommended that externally conducted investor presentations be posted on its website. This helps to strengthen the principle of equal access to information. These presentations will typically include investor presentations given by the company in connection with:

- Annual and quarterly reports.
- Investor meetings and conferences.
- Theme-based ad hoc presentations organised by, e.g. brokers.
- Webcasts.
- CMDs.
- Annual general meeting (AGM).

Industry and Sector Information

As discussed earlier in the chapter, the trend for listed companies is to provide greater information about their markets, competitive conditions and the industry in question. It is our conviction that that development needs to both continue and be supported. Except for business secrets which the competition may use adversely, there is no reason to seek to conceal these matters. The best equity analysts will still uncover the real conditions, and this knowledge will be disseminated through research report or verbal dialogues with investors.

The advantage of disseminating this knowledge is that it achieves a more transparent financial markets and thus more efficient pricing of its shares because more market participants have access to the knowledge in question. At the same time, this reduces the risk that part of the market has a higher level of knowledge than others. It is in the interest of everyone, and especially the interest of retail investors. This also reduces uncertainty regarding the company and, all other things being equal, leads to a lower risk premium on the company's shares, and thus a higher share price.

In our experience, far too many companies' senior managers are silent about matters which they consider to be confidential and sensitive from a competitive point of view. However, the truth is often that these companies are doing themselves and the company's shareholders a disservice.

By voluntarily providing the information in question, the company and its management obtain significant goodwill with investors and equity analysts, goodwill and trustworthiness that one may need at some later point in time.

Meetings with Investors and Equity Analysts

The purpose of meetings with the financial community is evident. They serve as the best marketing tool to attract new investors and keep current investors interested. They create a dialogue related to the perception of the company's standing in the financial community. A company typically meets investors and equity analysts at different kinds of meetings.

- Group meetings, where several investors and shareholders participate (from two to 100 + people).
- One-on-one meetings are with one or more people from the same investor or broker.
- Roadshows (where management or the IRO travels and visits investors and equity analysts).
- CMD (where management invites investors and equity analysts to a thorough update of the company's situation, typically a half or a whole day for larger companies).

There are several advantages and disadvantages to the group and one-on-one meetings.

The group meetings are rational. The company gets its message across to a larger circle at once. Some investors prefer group meetings if they do not know the company very well. They thereby benefit from the knowledge and questions from the meeting participants who know the company well. The very well-informed investors often only attend these meetings to ensure they do not miss anything or hold no one-on-one meetings. The company's one-to-one meetings to continue annual or financial statements are typically initially reserved for investors, not equity analysts.

One-on-one meetings are time-consuming. A company only meets one institutional investor at a time, but that is the only way to build good long-term relationships.

In other words, both group meetings and one-on-one meetings are necessary ingredients in the toolbox of the IR function. But everything at its time. Of course, it is different how many annual groups and one-on-one meetings companies hold, but it is not unusual for the larger company to hold more than 250 annual meetings per year, all inclusive. This figure is much lower for SMEs. So now we look at some of the other forms of meeting a company can use.

Roadshows

Domestic or international roadshows, where the company's management meets several investors and equity analysts and gives a presentation of the company, are held regularly. This is simply necessary to ensure that the company's investors and equity analysts are kept up to date with the company's development and regularly can meet the company's management.

Business journalists often seek permission with the company's IRO or communications manager to participate in the companies' investor presentations. However, in step with the spread of webcasts, in which business journalists most often listen in, a more liberal attitude exists today about business journalists' participation in investor presentations. Today, it is the main rule that business journalists are invited to significant investor presentations, e.g. in continuation of annual and quarterly results. In contrast, presentations to investors are often held in a small circle consisting exclusively of investors.

Journalists are typically not welcome to the latter investor presentations as the brokers, who are often behind these more exclusive meetings, want to limit the participants to only their customers. Systematised roadshows are generally carried out in continuation of:

- The annual report.
- Quarterly reports.
- Major mergers or M&A transactions.
- Major share capital increases.
- Major product or R&D events.
- Significant strategic changes or organisational changes.
- IPOs.

The individual company must assess both the frequency and scope of these roadshows. Again, this relates to the ambition level in the IR area.

We believe that a listed company should conduct a roadshow in continuation of both the annual report and the half-yearly report. At least one member of senior management should attend these roadshows. In addition, however, the IRO is often responsible for a possible roadshow in continuation of the accounts for the first and third quarters, respectively. The company should also pay attention to visiting investors and equity analysts appropriately.

However, with the advancements of video technology, more roadshows are conducted virtually. That saves time, but both the company and investors have a need to meet physically. That is in particular very important when building up a relationship with new investors and equity analysts.

In connection with the choice of cities to be visited, this to a very large extent depends on the company's listing and business focus.

There are three options for how the company can put together the visitor programme in each country:

- A broker arranges the meetings.
- An IR agency arranges the meetings.
- The company arranges the meetings itself.

The third solution is very time-consuming, and the company may overlook relevant investors. The second option costs money but may be a worthy supplement if the company is keen, or in need, to build an international investor base, or if no brokers wish to service the company due to a limited commission potential. The third option is the preferred one, if possible.

However, it is recommended that:

- The company itself actively participates with input from investors and equity analysts as to who it wishes to meet.
- Insist that equity analysts from competing brokers are also invited to the group meetings. The equity analysts help promote the company's shares via the equity analyst's research.
- The company does not use the same broker every time—competition between the brokers is good to obtain the best service. Furthermore, not all brokers have good close relationships with all potential investors—although they will often claim this themselves.
- Avoid blocking brokers because their equity analyst may negatively recommend its shares. These "wars" between the company and a negative equity analyst are destructive to both the company's reputation and the equity analyst's working conditions. A company has no interest in an equity analyst becoming mad at the company. As with a journalist, an equity analyst can inflict significant image damage on the company.

In contrast to ad hoc meetings with investors and equity analysts, it is recommended that the company prepare an actual script for its use at larger group meetings in connection with formal roadshows. This exercise ensures that the company provides the same consistent communication to all investors. Furthermore, it is also a great help in keeping time when more presentation slides need to be reviewed. Finally, of course, the manuscript should not be followed slavishly in the form of reading aloud but memorised for the most part.

Webcasts

Typically, the financial community prefers specific themes and not a long introduction of the company. Senior management should, therefore, primarily focus on new information. The financial community is seeking to gain an:

- Overall impression.
- Impressions of the management.
- Relevance.
- Discussion of the areas of interest to investors and equity analysts.
- Willingness to answer questions satisfactorily.

Teleconferencing or video meetings (e.g. MS Teams or Zoom) is an effective form of communication. However, it is even more important than with stock exchange announcements that the topic is material. A conference call or webcast typically serves the purpose of elaborating on the information contained in a stock exchange announcement earlier in the day.

In connection with telephone conference calls and any other form of dialogue with investors and equity analysts, the company must be aware that there is no potentially price-affecting information. A stock exchange announcement must be issued no later than the conference call's start if this is planned.

An invitation to a telephone conference is typically sent directly to the investors and equity analysts on the company's mailing list or may be contained in a company announcement. The invitation states if a PowerPoint presentation will be available on the company's website one hour before the start of the conference call. It is generally recommended to leave a PowerPoint presentation available. Finally, the invitation to the conference call should include several standard information, including:

- Date and time of the conference call indicating the time of different time zones.
- The topic of the conference call but be aware not to include price-sensitive conditions.
- Overview of company representatives who will participate in the webcast.

- Link, or telephone conference telephone number depending on the country where participants live.
- The link or conference call phone number to hear a recorded version of the conference call within 48 hours.
- IR's contact information.

In connection with longer telephone conference calls, the company should arrange a transcript, i.e. a written reproduction, of the conference call and have it posted on its website together with the PowerPoint presentation. It is a good service to users, investors, equity analysts and journalists, but of course, it costs both money and time, as the transcript usually must be checked and edited. On the other hand, the advantage for investors and equity analysts is they made review the transcript at a later time.

Especially during longer telephone conference calls, preparing an actual manuscript for the senior management's use is recommended. It is necessary to be very precise in its communication and fit the time. Investors' and equity analysts' patience is usually short for conference calls. Until the company reaches the final Q&A part, the communication consists exclusively of one-way communication from the management. The duration of the company's presentation—excluding–Q&A—should, depending on the company's size and level of activity, not exceed 30 minutes and rather be around 20 minutes.

Capital Market Day (CMD)

A CMD (Capital Market Day), a ESG day or a R&D Day (primarily pharmaceuticals and biotech sector) should be held much less frequently than webcasts and telephone conference calls. At most once a year or rather once every two years. If the company holds a CMD, it wants to provide an update with meat on, e.g. a thorough review of:

- New products.
- One or more of the major geographic markets.
- The company's research and development portfolio.
- A new or revised strategy.
- A major merger.
- Significant matters relating to the company have changed.

It is also often seen that a company issues a company announcement in connection with the beginning of a CMD with potentially price-affecting information that the company intends to disclose during the day. A CMD is not a traditional investor meeting in continuation of an accounting announcement. Instead, it is an investor meeting or an investor day, inviting both investors and equity analysts. The day is based on a theme or similar.

Before the CMD, the company must decide whether the CMD has the substance to last two, three, four hours or a full day.

A manuscript prepared in advance is also recommended for connection with CMDs. It provides the good discipline to perform this exercise.

The company must be ensured that it does not hold its CMD on the same day as one of its peer companies. It would be a pity to see a reduced attendance for that reason. Especially in connection with CMDs, it is recommended that companies less familiar with the area seek inspiration and sparring in the planning and implementation of the CMD.

The Company's IR Website

Constant involvement within IR and best practice, and as such, we will provide ideas and considerations. Therefore, our main advice is to seek inspiration with sector leaders who develop their IR tools, include mapping of industry peers and track the ongoing development.

This is one of the most important IR work tools—it must be updated and focused on the different relevant stakeholders. Include equity analysts, full-year guidance, new and OLD reports transparent. All the below. Product information, calendar, etc. When increasing focus, two options: (1) external assistance with mapping websites in the industry, or (2) internal assistance, review of closest peers and industry leaders.

- Overall impression.
- Ease of navigation.
- Speed of use.
- Equity story is clear.
- Financial goals are clear.
- Other objectives and policies are clear.
- Availability of relevant information about the company.
- Default information appears, e.g. board of directors, senior management, articles of association, corporate governance-related matters, IR policies, other policies, etc.

A company's website is the most important information channel for the majority of the company's users, which, in addition to investors and equity analysts, includes all the company's different stakeholders.

It is not the objective of this book to come up with a complete recipe for the structure, content and design of the relevant website, including IR website. However, it need only be stated here that the website's final design, and in particular content, is very important.

The availability, navigation and presentation of the information is crucial to retaining the user, and it is therefore strongly recommended that the company allies itself with the right expertise both in the critical initial stages of the website's design and coding/construction, as well as its later ongoing maintenance. Rapid development is taking place in the field. New possibilities and functionalities are constantly being developed. That said, both the IRO and communications manager can go a long way by studying how other best-practice companies approach the matter—not to emulate, but to learn, seek inspiration and hopefully take it to the next level.

Benchmarking of IR Relative to the Company's Peers

There are different companies and agencies rating companies' IR functions. However, the best input and feedback are typically obtained via own efforts.

If the IRO has a close relationship with investors and equity analysts, they have an ongoing opportunity to obtain valuable input regarding the IR area's performance. Of course, it is hardly delivered completely unfiltered by the investor or the equity analyst, but it is still input, and user input may be obtained from several different people. And then it's free.

The alternative, and much preferred approach, is using an external IR consultant who can carry out a so-called shareholder perception study. A perception study is based on interviews among IR's stakeholders, i.e. investors and equity analysts who follow the company. Further, it all about finding a competent and experienced IR consultant who the stakeholders will respect as the stakeholders will trust the IR advisor with private, and often sensitive, personal views. The stakeholders must feel that they are talking to a person who partly knows the company and its current situation well and partly has an in-depth knowledge of the premises under which investors or equity analysts work. If the company cannot choose the right IR consultant, time and money are wasted. It can also lead to bad will among the surveyed investors and equity analysts. It should be impossible to conclude the survey results with an adequate degree of certainty.

There are different rating studies of the IR functions of different internet-based companies, which independent firms and organisations publish from time to time, and they are a good temperature measurement. But they are not very useful for figuring out where the company can improve and what and what is really bad—often, this kind of studies only provides a relative measure to peer companies.

In the same way that the company benchmarks employee satisfaction, consumer satisfaction, products, distribution, etc., it is recommended that the company periodically benchmark itself in terms of actual performance, and against peer companies' IR activities. In conclusion, the purpose is to improve the company's IR tools.

Evaluation of IR Function

In connection with evaluating the company's IR function, one must examine the areas that investors and equity analysts attach the greatest importance. Therefore, several relevant questions to ask investors and equity analysts during an evaluation are posted in the following.

These are questions that the IRO can occasionally ask selected investors and equity analysts themselves if the IRO has a mutually trusting relationship with them. Since the questions in many cases are directly related to individuals, one should not necessarily expect a completely open and honest answer to all of them. Nevertheless, the answers will typically provide a good guideline if three or four or more of the most committed investors and equity analysts to express their views. Alternatively, or as a supplement, the broker typically also asks for feedback from the investors, either in general or in continuation, e.g. the company's investor presentations. The broker provides this service as part of arranging the roadshows.

However, it is recommended, e.g. once every two years, to have an external high-quality IR consultant (see "Use of IR consultants") to conduct an interview survey (generally referred to in the financial industry as a perception study) to obtain constructive feedback based on how the company is doing in terms of conduction its IR, and it may improve its IR function.

Suppose the company chooses to have an IR consultant conduct a perception study. In that case, it is recommended that the company, before the IR consultant is selected the investors and equity analysts to participate in the perception study, informs the individuals in writing that:

- The company has specifically selected the relevant investors and equity analysts. They will be contacted in a specific week by a specific consultant and asked several questions about the company and its IR function. This briefing should include a brief introduction to the IR consultant—both the firm and the individual in charge of the project.
- No individual answers will not be visible in the company's report from the IR consultant.
- The perception study is conducted to improve the company's communication with and financial markets service.
- The company would appreciate it if the investors and equity analysts in question would participate in the interview survey.

This approach creates an incentive for the participants to participate in the perception study, which increases the probability of a useful result of the perception study.

Finally, the questions should be sent to the respondents (the selected investors and equity analysts) to prepare in advance. This gives a better result of the perception study.

This approach before the perception study illustrates to the investors and shareholders concerned that they have been personally selected. At the same time, they are also stimulated to take the perception study seriously and spend the necessary time on it. The latter is far from a matter of course, as investors and equity analysts work in a very hectic environment with constant competition for their attention. The various feedback from the respondents may be provided either in writing, or verbally via interviews. That is just one of the many decisions that have to be made by the company, advised by the IR consultant.

This approach increases the probability of a usable result of the company's time and money in connection with the perception study. The evaluation areas could thus include the following and would typically be judged on a scale from 1 to 5 (where five is best).

Use of IR Consultants

The use of external IR consultants is comparable to most other areas where external consultants can take place. However, using an inexperienced IR advisor without a background as either an IRO, equity analyst or institutional investor cannot be recommended for good reasons (unless the IR is very seasoned). These have not lived and breathed in the environment and can therefore not be expected to appropriately challenge, or bring any value, to senior management or an IRO of a listed company.

The use of an IR consultant to perform its day-to-day IR work is typically more seen in the Anglian Saxon markets, e.g. the US and the UK, than in the Nordics and Continental Europe, where the financial markets prefer to deal with the company itself (be that the CEO, the CFO or the IRO) rather an external IR consultant. However, that is a geographical market preference.

However, an experienced IR consultant, with roots in the financial community, can be very valuable as an inspirer and sparring partner for both the company's management and IRO in the role of challenger and the devil's advocate in connection with:

- Preparation of annual report—discussion of equity story and structure.
- Quarterly reports—follow-up in the areas that interest the financial markets.
- Other company announcements—the right presentation of the story.
- Crisis communication—damage control.
- Input to or preparation of investor presentations—structure and themes.
- Preparation of webcasts and telephone conference call—plan, themes and critique of a manuscript, and a rehearsal.
- Preparation of CMD—agenda and themes, manuscript critique and rehearsal.
- Sparring for and preparation of Q&A.
- Assessment of whether a piece of the given information is potentially price-affecting. This will generally occur in collaboration with the company's internal or external legal counsel.
- Shareholder perception studies (interview survey of investors and equity analysts)—ensure that they are professionally organised.

Naturally, hiring an external IR consultant is associated with a cost. On the other hand, the company typically has only one opportunity versus the financial markets when important matters need to be communicated, and errors can be fatal. However, generally listed companies are recommended to handle the direct IR communication with investors and equity analysts themselves as this a good way to build relationships with both investors and equity analysts. However, using an external IR consultant to assist with IR-related materials, drafting, facilitating meetings, pressure testing, challenging, and consulting are naturally value-added and time-saving elements in respect to exercising IR.

If a company is considering the use of an external IR consultant, we recommend doing due diligence not only of the IR consultancy, but also in the specific allocated adviser. Assess the relevant person's experience in the field, ask for previously solved tasks and seek permission to contact other companies that have had a task solved by the individual, and the company. Preferably the same kind that is considered in the specific case. Although an IR consultant should operate with a high degree of confidentiality regarding previously solved tasks, it should be possible for the IR consultant to find one or two existing or former clients who are willing to serve as references.

But beware, some companies call themselves IR specialists without real in-depth experience in the field.

Perception Studies

Companies often do not receive the full extent of the market's view, even when in close dialogue with the market. Therefore, perception studies are, in our view, the most undervalued IR tool available to a company and are by most companies not fully utilised according to the substantial amount of valuable information they may provide. The objectives of perception studies include obtaining new valuable knowledge, possible areas of improvements as well as perceptions from investors and equity analysts concerning the company:

- Investor relations, communications and management functions.
- Operational, strategic and corporate governance matters.
- IR tools and activities.
- ESG reporting and objectives (something almost unheard of one or two decades ago) (Fig. 21.1).

A perception study is an investigation by an external IR consultant based on a written and/or oral survey. Equity analysts and investors tend not to give the same answers directly but may provide more honest feedback in an anonymised perception study. We have outlined that equity analysts may rely on maintaining a strong relationship with the company and will not necessarily be critical to their work. Investors are typically more transparent (Fig. 21.2).

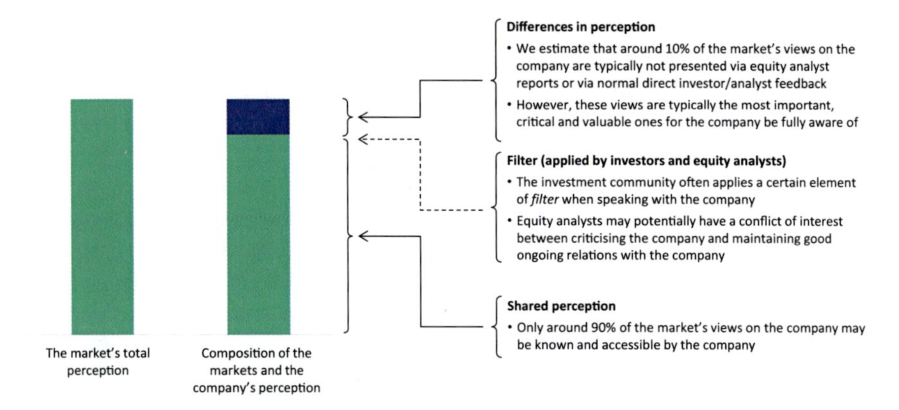

Fig. 21.1 A company's IR perception gap

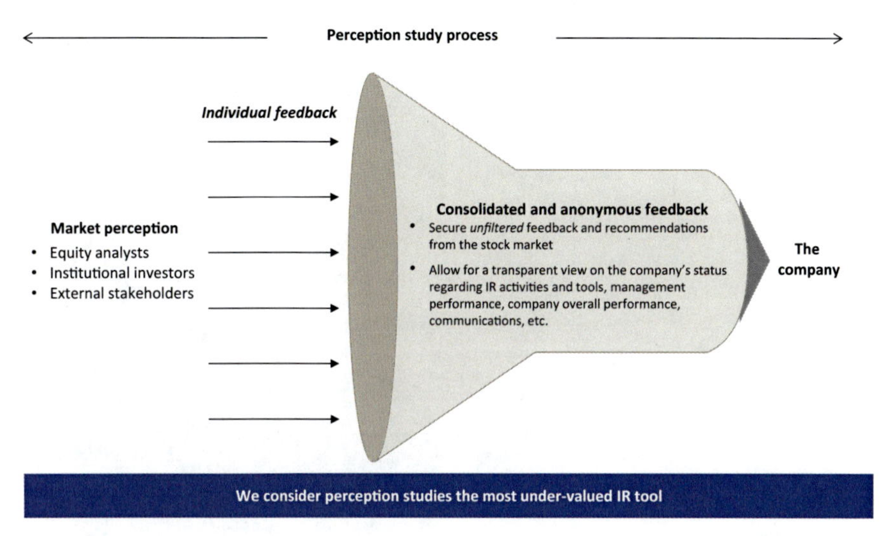

Fig. 21.2 Advantages of perception studies

Perception studies may be conducted either as an annual or bi-annual event to monitor the company's progress or in connection with various specific company events.

The full extent of the financial markets' views on the company is rarely provided through feedback alone. There is often a perception gap between how the company believes it is perceived and the full extent of how the financial markets perceives the company, as investors and equity analysts tend to hold back on the most severe criticism.

It is our experience that publicly listed companies often deem that they have a quite precise perception of the market's view on the company. However, this is in our view more likely to be true regarding only around 90% of the market's total views on the company. Typically, it is the residual approximately 10, which are the most important and valuable views, but which only emerge through personal, verbal conversations between equity analysts and institutional investors. Even the most forthright equity analysts and institutional investors typically apply some element of a filter when speaking with the company. The reason for this is, in our view, the existence of a potential conflict of interest between equity analysts and the company, as equity analysts typically wish to maintain good future relations with the company, e.g. for roadshow reasons, and to keep very informed by the company. This market feedback filter is eliminated through the completion of a confidential and anonymous perception study (Fig. 21.3).

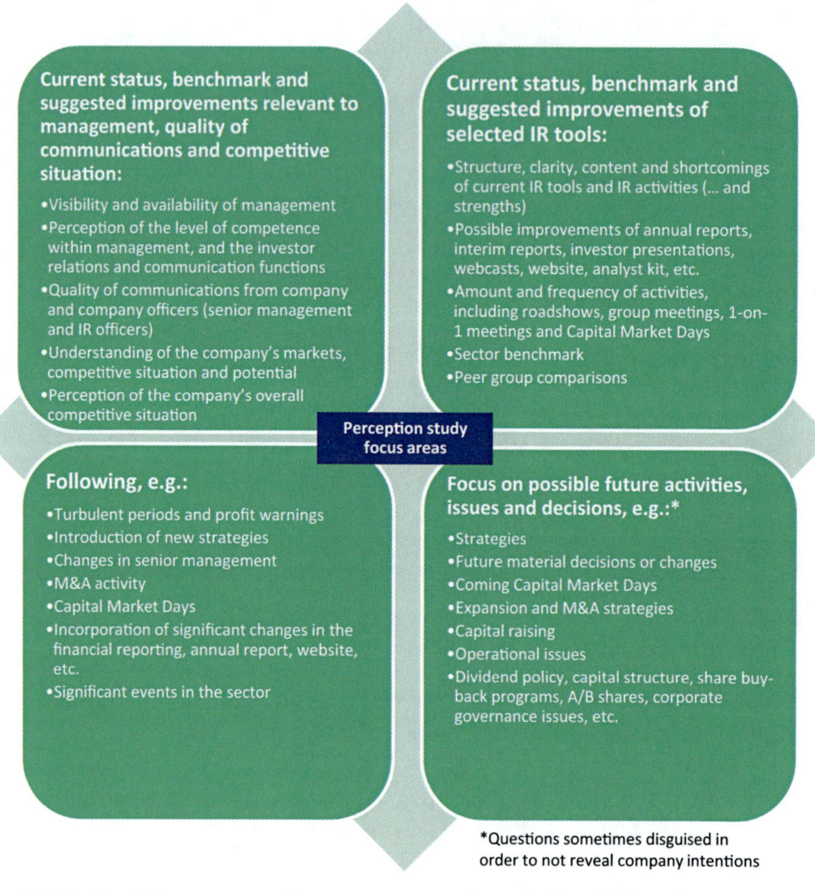

Fig. 21.3 Possible areas to focus on perception studies (*Questions sometimes disguised in order to not reveal company intentions)

A company should hire a professional external consultant to conduct a perception study. It should be a firm with strong and documented experience with the financial markets, including institutional investors and equity analysts. The higher credibility and trust the IR advisor has, the most honest feedback will be provided by the participants. The main questionnaire should be qualitative and quantitative and may include:

- A 1-page introduction of the external IR consultant.
- Inform that the study is anonymised, and answers are highly appreciated.
- Comments on senior management members, e.g. CEO and CFO.
- Commentary on individual IR representatives on:
- Detail of equity story.
- Areas of success and improvement.
- Major concerns.
- Financial reporting and objectives.
- Quality of the company's IR tools, including possible areas of improvements
- Corporate governance and board of directors' issues.
- ESG reporting and objectives.

The perception study results may subsequently be provided to the company in an anonymised PowerPoint presentation and/or a written report. If successfully conducted, the perception study results will be the best mirror on the company's IR-related communication. Therefore, it is important to take exercise seriously. It is important to be followed up with an internal seminar and work session to discuss and implement the feedback with the external advisor. The quantitative study is recommended to be followed up 1–2 years after the previous perception study to track the company's performance as to the quantitative ratings obtained in the perception study (Fig. 21.4).

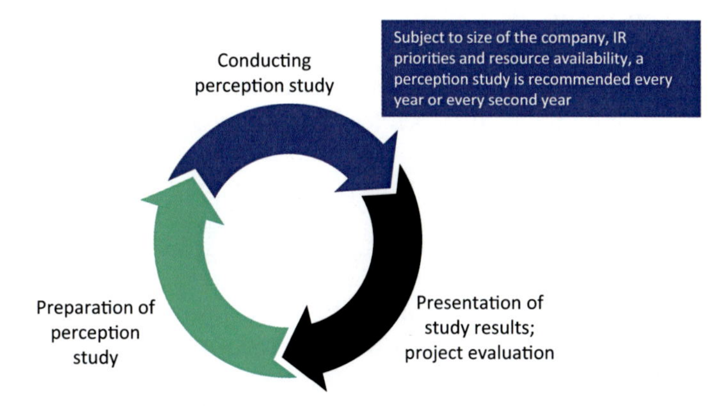

Fig. 21.4 Perception study—project cycle

Who Should Commission an IR Perception Study?

Typically, an IR perception study performed by an external IR consultant is nearly always commissioned by the IRO, the CFO or the CEO. Then, later when the perception study is conducted and a report or a presentation has been delivered to management, then the perception study report is—sometimes—provided to the board of directors. That is all also very well, but occasionally it happens that an IR consultant frequently used by the company's management over a period time has become close to the IRO and management to an extent that an external IR consultant perhaps is not too keen on delivering the most negative feedback from the respondents (i.e. the investors and the equity analysts) to the IRO and management. Further, there may also be a delicate situation, where the financial markets are negative on the board of directors, and that may be something that the IRO and management are not too keen on providing to the chairperson via a perception study commissioned by management. We are, ourselves, aware of incidents, where a negative perception study report did not even reach the CEO of a company (withheld by the IRO) due to negative comments about, yes, the CEO.

The bottom line is, however, that boards need to know exactly what investors and equity analysts are thinking.

Unfortunately, some boards do not pay sufficient attention to investor perceptions and do not actively request or receive this information from management. Even when management does have a good dialogue with the market, it is not guaranteed that investors' views will reach the board of directors in an unfiltered manner. Furthermore, management may often believe it has a complete picture of how the market feels about the company when this is not actually the case.

The level of insight boards are expected to have into their companies has increased considerably over the few past decade, and many events have shown that a failure to understand major shareholders' concerns can prove fateful for the board.

To perform optimally, the board needs direct, unfiltered access to the views and feedback from investors and equity analysts—not only on the company's business model, strategy, products, pipeline, reporting, capital structure and governance, but also unfiltered market feedback on the company's equity story as well as on the CEO, the CFO, the IRO and the board itself.

We consider investor perception studies to be the most underestimated yet powerful of IR tools. Investor perception studies are carried out by external IR experts who have not only an in-depth knowledge of the players in the market but should also have the trust of the selected respondents, meaning that significant criticisms can be made safe in the knowledge that they will be communicated completely anonymously to the company.

We expect the following five guidelines to become the norm among boards within a few years, as it currently is among the best practice international companies within IR, and which strives for best-practice governance and optimal stakeholder relations:

- IR should be a fixed item on the board's agenda, driven by the chairperson.
- The IR function should set quantitative targets for equity analyst coverage, share liquidity, shareholder mix and share price relative to fair value, along with instructions for reporting to the board—all challenged by the board of directors.
- External perception studies should be carried out (typically every year or two, depending on the size of the company).
- To ensure and validate the independence, the chairperson should select, or be involved in the selection of, the external IR consultant to carry out these studies, the selection of respondents (typically between 15 and 50 equity analysts and institutional investors internationally, depending on the size of the company) and the formulation of the scope of the questionnaire. There is little reason why the IRO and management of the company should not use one firm of IR consultant for its daily or occasional IR advisory need, and then the board of directors another firm of IR consultants for perception studies, where the latter firm of IR consultant is completely free of any potential conflicts of interests, e.g. ongoing advisory assignments, and may therefore act completely independently vs. the board of directors.
- The board should receive the written report and a supplementary oral presentation at least at the same time as senior management.

A board of directors not having good, unfiltered insight into how the stock market's views on the company and its board and management is fundamentally at a greater risk of failure than boards will a solid insight of the views of investors and equity analysts.

Digital IR Tools

Equity analysts and institutional investors are heavy users of financial service programmes such as Bloomberg, Thomson Reuters and FactSet. Equity analysts typically have the best access to data through their platforms and support from the brokers and even investment banks. As equity analysts benefit from having a strong relationship with the IRO, it is not uncommon for the IRO to ask for specific data on an ad hoc basis, which the equity analysts will happily provide. Therefore, the majority of IROs do not need to have access to the same financial service programmes as the investors or equity analysts.

The IRO needs to have an up-to-date view of the company's perception, information flow, share price and other critical financial data points. In addition, however, the IRO is often responsible for administering the current shareholders, managing internal option programmes and mapping the dialogue with investors and equity analysts. There are a range of services and software to assist on this. Below, we have named a few suggestions to get an IRO started:

- **Q4 CRM** (www.q4inc.com): A tool that provides the IRO with existing and target investor statistics. It has an investor targeting database with information on a fund level and a relevant contract person. The database is highly accurate as it collects data from 13F filings, Bloomberg, Thomson Reuters and FactSet. It also has add-ons that can provide details on nominee accounts based on phone surveys to investment funds conducted by the firm.
- **Computershare** (www.computershare.com)**, IHS Markit** (www.ihsmarkit.com/product/investor-relations-solutions) **and CM i2i** (www.cmi2i.com): It is helpful to have tools that track existing investors. However, it does not separate with nominee accounts as Q4 is able do. However, the tool better allows strategic intelligence regarding current and prospective shareholder identification. These are essentially platforms that assist the IRO within market intelligence, CRM optimisation, IR strategy, investor engagement, perception studies and ESG analysis.
- **Microsoft Navision** (https://dynamics.microsoft.com): The most widely reporting accounting programme which provides the IRO with relevant data published in a trading update. The IRO can download the information into excel and modify the data into the corporate design to be included in the public information.
- **Holding.se** (www.holding.se): A widely used and affordable platform in the Nordics that collects information from shareholders' public databases. It is mostly investment banks that use this system to receive a more insightful overview compared to several other financial platforms.
- **Nasdaq IR Insight (nasdaq.com/nasdaq-Ir-Insight)**: An all in-one-workflow insight system to maximise the effectiveness and value for IR programmes. The system offers investor insights and targeting statistics, market and peer Intelligence, and investor engagement tracking.
- **Euronext Securities (VP.dk)**: The CSD network recently acquired VP securities and renamed it Euronext securities. The tool is widely used in the Nordics and is especially useful to administer and manage employee option programmes. It can also gather statistics and materials on relevant shareholders. The tool is especially an effective compliance tool for the IR and dialogue mapping with investors and equity analysts.

CHAPTER 22

Managing the Expectations of the Financial Community

A Proactive Approach to Managing Expectations

The more transparent the management is, the lower risk of high volatility and deviation from the market consensus estimates the company will experience. The company should be aware of its competitive environment when making information public, but a high level of transparency, in general, portrays credibility and confidence on behalf of the company. Management should not provide detailed numbers on all topics but can provide ball-park numbers to underpin management's statements and thoughts. However, it is also worth noting that once the company has decided to communicate at a certain level of detail of information, it is difficult for a company to reduce its transparency later.

The IRO is respectful of equity and investors, including retail shareholders. They do not discriminate against the types of actors, even though their perspectives are critical. The company should never entrust confidential matters to investors or equity analysts and never reach negatively towards criticism. Further, it is essential never to comment on the share price and be forthcoming, also towards difficult questions. It is the companies to run the business, and the financial markets' task to evaluation and project the share price of a company.

Equity Analysts

Generally, equity analysts are respectful of the information they are given. However, the motivation is to find an information advantage and deviation from consensus in the communication with management. Therefore, any information communicated to the equity analyst can likely be passed onto

P. Lykkesfeldt and L. L. Kjaergaard, *Investor Relations and ESG Reporting in a Regulatory Perspective*, https://doi.org/10.1007/978-3-031-05800-4_22

investors—and potentially even the media. Therefore, it is essential to be prepared when talking to equity analysts in formal and informal settings.

Before a formal meeting, it is recommended for the company's representative to be updated on the equity analysts' view of the share, current recommendation and estimates, and target price deviation (from the company's internal confidential intrinsic valuation of its own share). However, instead of correcting an equity analysts' assumptions, one may point towards consensus, extraordinary items, market information and past commentary to guide the estimates. However, never in a too detailed manner. For example, if an equity analyst includes a CAPEX level of 500 million euro for next year, which the IRO finds too low. In that case, the IRO may remind the equity analyst of the company's current plans, and what consensus is expecting. There is a fine regulatory line, and it may be a careful balance guiding the market while at the same time not disclosing anything remotely being characterised as sensitive information.

Not all analyst reports are read by the IRO before publication. However, the equity analyst may want to publish an in-depth or initiation report and ask for commentary by the IRO. The IR will typically not receive the valuation, the front page, the investment case nor the investment recommendation or the target price—only the information relating to the market, financials and perhaps the equity analyst's estimates and assumptions. Like any other setting, the IRO may factually comment on the information and perhaps discuss selected assumption, but should avoid providing comments on the future, including any on forecast estimates, etc.

It is essential to be balanced and not be too optimistic or pessimistic. For example, the investment bank usually requires a four-eye principle or a compliance office to attend meetings where an equity analyst discusses non-published reports with the company. Likewise, the IRO could consider inviting external or internal councils. All with a view to avoid later regrettable misunderstandings.

Investors

Institutional investors and buy-side analysts tend to be less short-sighted than equity analysts, as the first group do not have to make publicly available three-year financial estimates. Institutional investors are more concerned with long-term strategic considerations than quarterly numbers. As a result, institutional investors require more insight into the long-term potential of products, competitive trends and the management's thoughts on HR, talent and ESG issues.

In general, institutional investors prefer to receive input from a range of equity analysts, the company and its peers. Therefore, they are less considerate of sharing new information back to these actors. Therefore, the IRO and management can be slightly more transparent on granularity towards the institutional investors and shareholders relative to the information distributed to the equity analysts—although no selected disclosure must ever take place.

Managing communication with retail investors may be difficult. The best way is to monitor their considerations on various social media platforms and focus on the company's core equity story. It is recommended for the IRO and management to be marginally more conservative when communicating to retail investors, as they tend to be less balanced when investing, i.e. either over-optimistic or over-pessimistic. The best way to manage expectations with retail investors is to maintain a balanced communication in the financial media, the main source of their information—other than the information made public by the company via the stock exchange.

CHAPTER 23

Embracing the Digital World of IR Activities

Transparency Is Increasing for Retail Investors

The technological revolution has allowed unprecedented availability of investor information, data and insights on a wide range of industries and companies. In addition, selected digital platforms facilitate contact between companies and retail investors. This means that a company may communicate directly to its retail shareholders. This has allowed companies to increase their shareholder communication towards retail investors like never before.

Companies have begun to embrace digital IR. It should be clear from this book that it is our view that the scope of IR is wide-ranging and that the clear purpose of good IR is to achieve a fair value of the share price. However, a "fair" value is not the same as a "high" value, and therefore, we raise a flag of caution in respect to any platform that may implicitly market itself as apparent independent "Digital IR" while it in fact may serve as a non-independent marketing platforms towards retail investors on behalf of selected companies, who in turn pay the digital platform for the marketing of its share. The digital platforms connecting companies and retail investors play an important and positive role in the financial markets in respect to retail investors. However, in the event that such digital platforms are facilitated by individuals employed by the digital platform company, these platforms must be deemed to be a marketing vehicle for the company and have the same built-in potential conflicts of interest as it is the case with commissioned research which is also paid for (or sponsored by), the company. When embracing digital IR, the IR department must have a clear purpose, goals and objectives to increase its exposure on social media platforms—but also be aware of any build in potential conflicts of interest.

P. Lykkesfeldt and L. L. Kjaergaard, *Investor Relations and ESG Reporting in a Regulatory Perspective*, https://doi.org/10.1007/978-3-031-05800-4_23

Applying Social Media in IR

As discussed in Chapter 2, professional market participants attempt to piece together information by finding data points and ideas, discussing them with other market participants, and they yet again piecing together the information, attempting to undercover an information advantage.

The same tendency is now possible for retail investors, who can communicate and discuss investment cases and ideas on social media platforms. More and more companies now embrace this tendency and also have an active approach to it. In addition to hosting a publicly available webcast, the IRO and management may consider discussing their company on social media platforms by conducting interviews or publishing short films. An increasing number of companies have begun to be live host sessions where retail investors can ask specific questions.

The trend is expected to continue to increase as social media and organised retail investors gain importance.

In addition, to have a comprehensive website, the IR department may coordinate with the in-house marketing function to align a company's social media strategy with its IR strategy. For example, as more investors follow trends on social media, sensitive information must not be shared on these platforms. If a new company announcement from the company is shared on social media, this creates strong synergies, as a user of a company's products may not necessarily know if the company shares are publicly listed. Such a consumer may with this new knowledge decide to invest in the share of the company whose products the consumer have a preference.

Potential Pitfalls of Digital IR

The field of digital IR remains new. During the past years, we have witnessed a situation where some CEOs of major companies have been too vocal on social media platforms causing unnecessary attention from the regulators and crowds' behaviour among retail investors. The IRO has policies and contingency plans when correcting mistakes in a controlled environment. However, information moves extremely fast on digital media and platforms. Therefore, if a company chooses to engage in social media, it is vital not to overshadow the effects of managing expectations towards institutional investors and to have proactive contingency plans in place if problems arise. Due to two-way communication, expectation managements and regulatory professionalism of equity analysts and institutional investors, it is recommended for most companies to prioritise their needs ahead of a company's IR strategy, and before potential problems emerge.

A summary of facts and best practices is illustrated on page 329.

PART V

IR in Special Situations

CHAPTER 24

Preparation of Difference Types of IR-Related Contingencies

> When everything seems to be going against you, remember that the airplane takes off against the wind, not with it.
>
> —Henry Ford, Founder and CEO of Ford Motor Company Inc.

A Crisis Communication Contingency (CCP)

We have earlier explored best-practice IR and the regular duties and working areas of the IRO. This section dissects how a company and its IRO can prepare and develop different types of contingencies and act in special situations. Special situations are considered unexpected events, like takeovers, crises and sector-specific situations. To prepare and mitigate for these situations, a company is recommended to develop a crisis communication contingency (CCP), an issue management contingency (IMC) and a takeover response manual (TRM) (historically known as "defence bibles"). For stock market-related events, the responsibility typically rests with the IRO, while for non-stock market-related events, the responsibility typically rests with the communications department.

For any type of communication, issue management and crisis contingencies, companies are recommended to identify the important issues that can potentially go wrong and may have serious consequences for the company. These issues may be both general and sector-specific and can relate to, e.g. fires, ship sinking, exploration, deaths, emissions or similar. A CCP is a communication game plan for internal and external communication. It includes several appropriate "drawer statements", i.e. several different drafts of internal

P. Lykkesfeldt and L. L. Kjaergaard, *Investor Relations and ESG Reporting in a Regulatory Perspective*, https://doi.org/10.1007/978-3-031-05800-4_24

announcements and press and stock exchange announcements. The statements should consider several different scenarios in connection with special situations.

The purpose of a CCP is to ensure proper and fair communication in a crisis that the company cannot avoid communicating about. The CCP may also try to define what constitutes a crisis and a timeline of actions to be taken. It may save valuable time instead of waiting for a green light from senior management. It must be able to be executed quickly and consists of two parts:

1. A part that defines what is to be said.
2. Who should do what and when—in the given scenario?

How to Design a CCP

A CCP is prepared well in advance of the crisis in the concept of "emergency preparedness". Therefore, it is important that emergency preparedness only contains activities that it is realistic to carry out in a very short time.

The emergency preparedness must contain the messages and information expected to be needed. In addition, the CCP must contain a generic action plan, which can be implemented from the moment the crisis occurs (Figs. 24.1 and 24.2).

The above reporting of status on issues provides a greater overview than a lengthy report. Consider initiating all status reporting on issues with such visualisation. It will immediately indicate where resources need to be put into dealing with an issue.

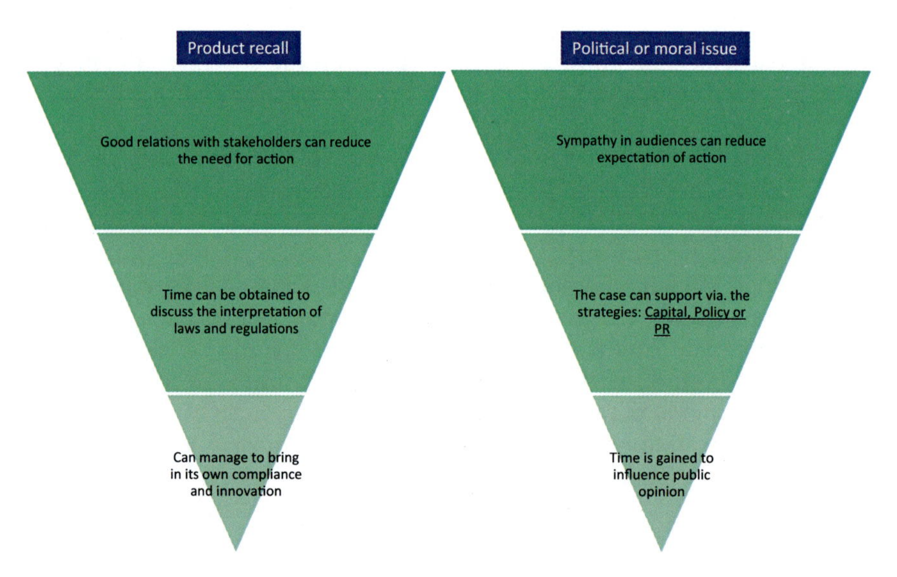

Fig. 24.1 Examples of practical issue management

Activities / phases	Phase 1, Stability	Phase 2, Emerging	Phase 3, Threatening
Reactive strategies	Stay quiet	Network with stakeholders and create alliances outside the stakeholders	Changes of policy and strategies
Proactive strategies	Create alliances outside the stakeholders	Launch of strategy: Capital, Policy or PR	Damage control

Issue	Stability					Emerging					Threatening					Action, deadline and responsibility
A	x															aa
B							x									bb
C													x			cc
D																dd

Fig. 24.2 Assistance in prioritising activities in a contingency plan

Formulation of Messages

In the event of a crisis for which the company itself is to blame or has co-responsibility, the company must try to gain control of the story. It can achieve this by being the first to communicate about the crisis and thus accustom the media and the public that the company can handle the crisis professionally. The company must therefore be the most interesting source of further information.

When the company itself is the first to communicate, it usually has free speaking time to tell what it wants to do to resolve the crisis. The messages must at least contain answers to the following questions:

- What has happened; where, when and how (to the extent possible to answer)?
- What is the consequence for the company and other stakeholders?
- What will the company do to stop the crisis?
- When is the crisis over?
- What will the company do to compensate for the crisis?

The last question is hypothetical but good to think about, as crises that get attention in public are quickly described as good or bad morals—is the company evil or good? Here, the company must improve some specific conditions that the public will perceive as goodwill. Compensation can be many things, from an internal investigation that temporarily pushes the case to the corner to an investment in the local community.

Action Plan

An action plan must define when the various activities are carried out. As a rule of thumb, the more the crisis is about a concrete event, which can have negative consequences for people, the environment and equipment, the faster communication must take place. The more the crisis is about administrative or legal issues, e.g. whether the company has committed an administrative error that is not widely known to the public, the more appropriate it is to communicate about the matter.

There are also crises that the company does not need to communicate publicly; for example, it is best practice for pharmaceutical companies to proactively communicate when withdrawing a product from the market, even though this information is publicly available.

The Employees

In the event of a crisis, communication with employees is essential. In a crisis, the management must seriously show that it is worth the employees' trust. Therefore, it can be beneficial if the leaders across the organisation set aside time to talk to employees about the crisis so that it does not become taboo afterwards. One can invite a representative from the media to such meetings to focus on the company taking the crisis seriously internally. All broad communication towards the employees is considered public information, as this information tends to spread quickly.

Use and Maintenance of Emergency Preparedness

Once the contingency plan has been formulated and approved, it should be reviewed once or twice to document that the established schedules can be met. Otherwise, the contingency must not be used in any other way than that it must be available in an updated form when—perhaps—it is needed.

Internal Workshops and Crisis Simulation

It can seem overwhelming to rehearse a contingency on every single scenario that may only occur. Especially if it is a case that is not directly threatening to the company. However, best-in-class companies do execute fire drill exercises on a regular basis, facilitated either in-house or by external advisers. Therefore, the company may instead decide to rehearse some general situations, which must be in order in all crises and to test emergency preparedness. It may be that:

- Everyone on the roster can be met at any time and be available within an hour.
- The media trainer is available and can show up within the agreed time.

- Relevant professionals are available for further information.
- There is a connection to the company's network, and the IT supporter can be reached outside working hours if necessary.

Furthermore, it can be an advantage to establish relations with the relevant authorities, which the company knows will come into action if emergency preparedness must be taken into use.

CHAPTER 25

Developing a Takeover Response Manual

Preparing for a Potential Takeover Situation

Takeovers are widely used in the financial markets. Back in the 1980s, unsolicited offers on companies were referred to as "hostile" takeovers or bids. However, where is the hostile element? Hostile to shareholders? Hardly, in the case of a so-called full bid, which is an offer on the company's shares that fully values the company's short- and long-term price potential and synergy considerations—perhaps at a premium of 25–50%, sometimes more, of the share price prior to the announcement of the offer.

However, an offer may be seen as hostile towards selected members of the board of directors, senior management or the staff, respectively, and only if the takeover is expected to lead to significant savings measures. When a bid is high enough, and there is no white knight, i.e. a buyer that the target company (the company to which the purchase offer is directed) finds "friendly", then the offer becomes at some point friendly. This is because, by the series of current rules and recommendations, the board of directors is ultimately obliged to recommend the tender offer if the price is deemed to be sufficiently high.

For most companies, a takeover is the special situation most likely to occur. The work of preparing an IMC for a takeover can vary in scope, and it is up to each company to decide how extensive and resource-intensive the project work in connection with the preparation of an IMC must be. There is, of course, the "all-mentioned nothing-forgotten" version, but the company can also obtain a lot with a more limited but very focused work.

P. Lykkesfeldt and L. L. Kjaergaard, *Investor Relations and ESG Reporting in a Regulatory Perspective*, https://doi.org/10.1007/978-3-031-05800-4_25

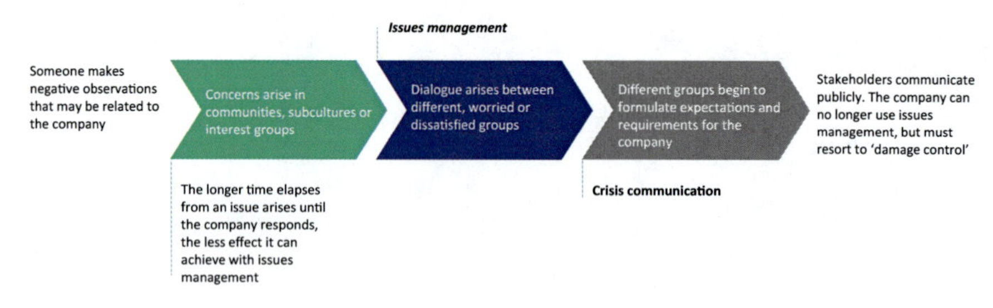

Fig. 25.1 Development of an issue (1/2)

An Issues Management Contingency (IMC)

The purpose of an IMC is to mitigate potential issues from arising that could harm the company. An issue arises when there is a gap between the stakeholders' expectations of the company and the company's policies, performance, products or general obligations to the public. IMC is the process required to close this gap and can use many different disciplines depending on which issue is to be handled. IMC can use PR, lobbying, media relations, trend spotting, strategic and financial planning, law, etc.; however, it is not about:

- One-way control of political cases.
- Spin or damage control or crisis communication.
- Deliberate process delays or diversionary manoeuvres to eradicate opponents.
- Branding.

Preparing an IMC for different important events, which may significantly influence the company, is in the interests of everyone in the work group list along with external stakeholders. It is, therefore, imperative for the IRO to collaborate with the company's work group and identify potential issues and how they can develop (Figs. 25.1 and 25.2).

The company should develop an IMC for each identified potential issue with a fact sheet on each issue and Q&As. The company should also practice these plans with fire drill exercises as the issue may arise. In addition, standard draft statements of hypothetical scenarios, fact sheets, etc., should be up to date if a situation arises (Fig. 25.3).

How to Design an IMC

An IMC may be designed in many ways depending on the issue and local conditions. In any case, a contingency plan must consider what can be done so that the implementation of the IMC is subsequently perceived as legitimate in the public eye.

Fig. 25.2 Development of an issue (2/2) (*Audiences may impact the company with their views without being immediate stakeholders)

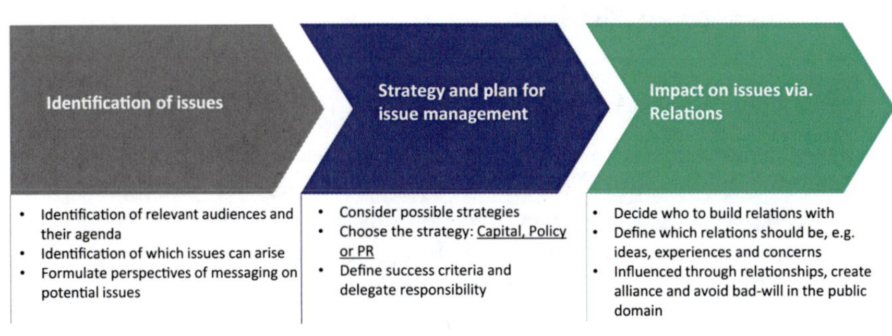

Fig. 25.3 Practical issue management

Suppose the company needs to oversee many issues at different locations. In that case, it may be considered to establish a web-based reporting system where local managers can report incidents and trends that may indicate that an issue is underway.

Identification of Issues

Identification of issues mainly consists of three steps:

- **First**: Assess current social, political and economic trends, including macroeconomic conditions and technological development.

- **Second**: Comparison of these trends with the company's strategic goals.
- **Third**: Identify what issues may arise in potential gaps between these trends and the company's goals.

If the company does not start by assessing the macroeconomic trends in a helicopter perspective, then the company risks only identifying the issues that the company is already aware of, and which should already be identified in the company's risk management systems and issues management contingencies.

Analysis of issues

The analysis should focus on what experiences the company and others have had with similar issues in the past.

What measures were taken, and can these inspire the company's current strategy? Use a quantitative and qualitative analysis of stakeholders' views on the current or similar issue.

Next, it should be assessed how the issue in question can affect the company, e.g. in three to five degrees of development. Then, finally, provide estimates for financial and organisational consequences and the company's future' license to operate in society.

Strategy for Changing an Issue to a Non-issue

Strategy is a strange word to use when it comes to issues management. However, the strategy must first and foremost determine the company's position on the specific issue. Then, it can be considered whether to "lie low in the grass" and let others take the rubbish, e.g. a competitor representing the industry. In that case, the strategy and the subsequent action plan must be reactive, and one can try to keep up with the flow (Fig. 25.4).

If it is assessed that a proactive strategy must be adopted, proactive work must be done to create an understanding of the company's position; careful consideration must be given to how this understanding can be built up.

The following three types of initiatives can generally outline a proactive strategy for creating an understanding of the company's position:

- **Money**: Assess whether it can help financially support someone, an individual or an organisation. In the US, it is not uncommon for a pollution issue to be followed up by a donation to an environmental organisation that accepts this approach.
- **Politics**: Assess whether it can help create "political results" in the local community. Examples include creating new jobs, paying taxes, offering education, social benefits, etc. If the purpose is noble enough, there are very wide limits to how much one can afford to invest here.

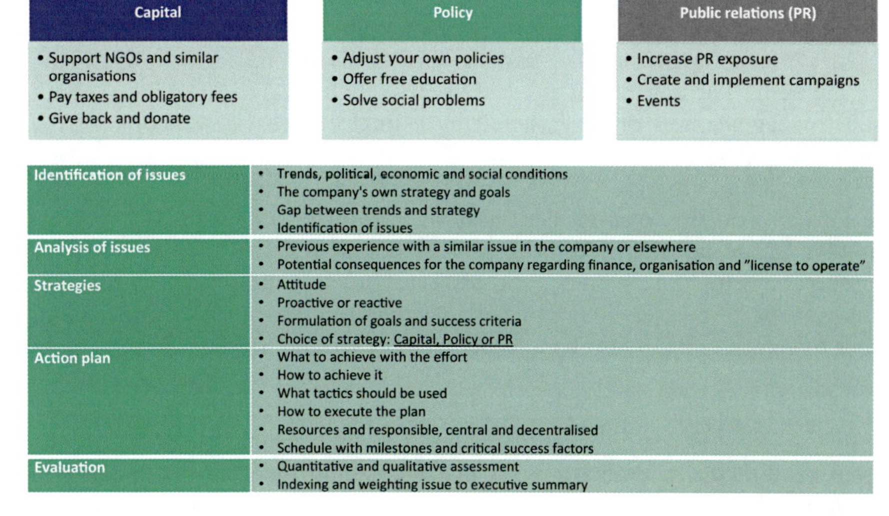

Fig. 25.4 Three types of issue management strategies

- **PR**: Assess whether it can help to communicate with selected groups at personal meetings, via the media or events. The company cannot communicate an issue via the media, but the media can be an important ally if an issue arises. Therefore, make sure that those who can write about the company have a thorough understanding of and acceptance of the company's work.

Action Plan

When formulating a plan for dealing with issues, it can be tempting to start by establishing relationships with the stakeholders who need to be affected. But before that—because the company must get established and utilise the right relationships—the company should formulate:

- What to achieve with the effort?
- How to achieve it?
- What tactics should be used?
- How to execute the plan?

Then, it must be determined whether any resources in the company must be involved. Of course, it would be ideal if everyone who can naturally contact relevant stakeholders and key opinion leaders were involved and given a role in emergency preparedness. However, this approach will often encounter the habits and organisational conditions that the company operates according to everyday life.

Therefore, it may be helpful to set up a task force with direct reference to senior management, who is responsible for handling the issue.

Does the company need a person who is always responsible for issues management? The question is relevant because the necessary knowledge about issues management is often gathered by a single or very few people in the company.

The optimal situation would be that such a person is responsible for the formal processes around identification, formulation of plan and strategy, but that the responsibility for the process lies with the BU that represents the current issue. Alternatively, the responsibility can be placed with senior management. However, it is not realistic that the person who understands issues management can also handle all activities around a single issue.

Evaluation of Results

It can be difficult to evaluate issues management, which may extend over a very long period. This is because the specific issue changes character, and the people who must deal with it are constantly replaced.

Therefore, the action plan must contain some concrete goals and critical success factors, which can be assessed objectively over a very long period. Indicate if necessary. With a weighting of the degree to which it has succeeded in meeting goals and critical success factors and let an overall weighting make it out of the issue's status in question.

That way, it is relatively simple to make an executive summary of all the company's issues, listing each issue on a line with a number indicator of how things are going and how big the risk is of the issue breaking out. Combine if necessary. With a red-yellow-green colour scale in the weighting. If the overview is outlined as simple as described here, it will be a welcome status at company board meetings once a month or during the quarter.

Use and Maintenance of an Issues' Management Contingency

An IMC should serve as a living project description, where everyone involved is constantly informed about adjustments in the plan. Assess whether the emergency preparedness can be stored electronically with selective access for the persons involved.

An IMC must be continuously updated. This is because, first and foremost, an issue is constantly changing, but also because the company's goals can change in the short and long term to influence its attitude to an issue. Therefore, it is recommended that an IMC be updated at least every three or six months if management is regularly briefed on the company's issues. It is the task of the company's issues management manager to ensure that the emergency preparedness is continuously updated. It may also be appropriate for the person in charge to help the senior management to ensure that the planned activities are carried out.

A Takeover Response Manual (TRM)

These considerations need to be taken into consideration when constructing a TRM. The TRM should be headed by the CFO and include the IRO, communications manager and external financial advisers. If a takeover bid is made, a well-planned TRM will make life somewhat easier for the board of directors and senior management to make the appropriate decisions. A well-planned TRM also means that:

- The probability of rejecting the purchase offer is optimised in the event of an insufficient bid.
- A higher price is typically obtained for the company's shares if the company is ultimately taken over.
- During the siege, the company's everyday management and operations can be maintained as far as possible.

It is further assumed that the probability of a listed company receiving a direct or indirect inquiry regarding a takeover, or receiving an unsolicited offer for its shares, has increased in recent years, as:

- Company acquisitions are increasingly included in companies' expansion strategy due to investors' demands for an optimal capital structure, resulting in a higher debt ratio.
- Investors are increasingly uncompromising in their pursuit of higher returns.
- The private equity funds have more capital available than ever before.
- International interest rates are historically low.
- A reduction in foundation ownership has occurred in some listed companies, either due to the need for new capital or to modernise a given company's corporate and ownership structure (Figs. 25.5 and 25.6).

The Best Protection Against a Takeover

It should be stated that the best long-term and most future-proof protection against a takeover or a takeover attempt is a price on the company's shares, which fully reflects the company's both short- and long-term value potential, i.e. where the risk premium is as low as possible (Fig. 25.7).

If the company's both short- and long-term value potential is reflected in the share price, there is less to attract third-party potential bidders. In any case, this should eliminate the value gap that financial investors and private equity funds, or peer companies, are in some cases able to take advantage of. It is

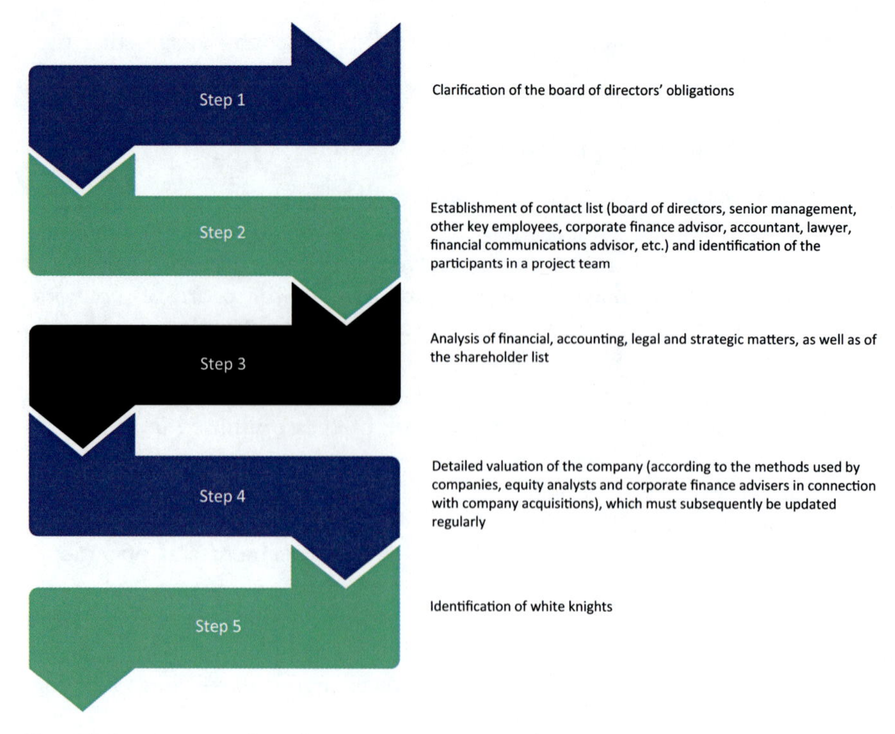

Fig. 25.5 Content of a takeover response manual

more difficult with the possible industrial synergies that a larger competitor can realise in connection with a takeover. However, there is only one thing to do: run as efficient a business as possible.

As to a possible takeover, it is not the intention of making a company's takeover difficult or impossible. On the contrary, good IR and good corporate governance mean that the company's information policy and ownership and management structures are transparent. Our clear position is that the interests of shareholders must first and foremost be safeguarded, to run the business as efficient in the longer terms and to reduce the risk premium as much as possible.

If the interests of the shareholders can be safeguarded by the management being able to ensure such a consistently high share price that a takeover of the company by a third party is financially uninteresting, this must be preferable. In such a case, it will be an excellent and well-run company. However, it does not go unnoticed that a takeover has high costs. The advisers cost a lot of money; the employees do not keep the same focus on the customers during the takeover period, etc. A takeover situation, successful or not, requires significant resources for senior management.

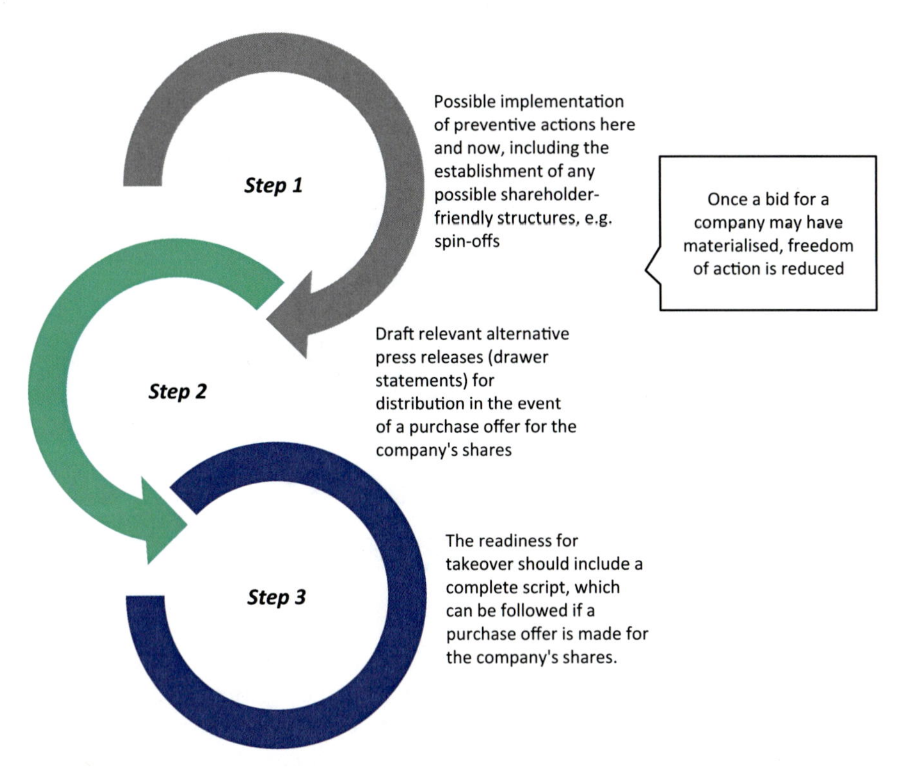

Fig. 25.6 *Content of takeover* a takeover response manual—*following a bid*

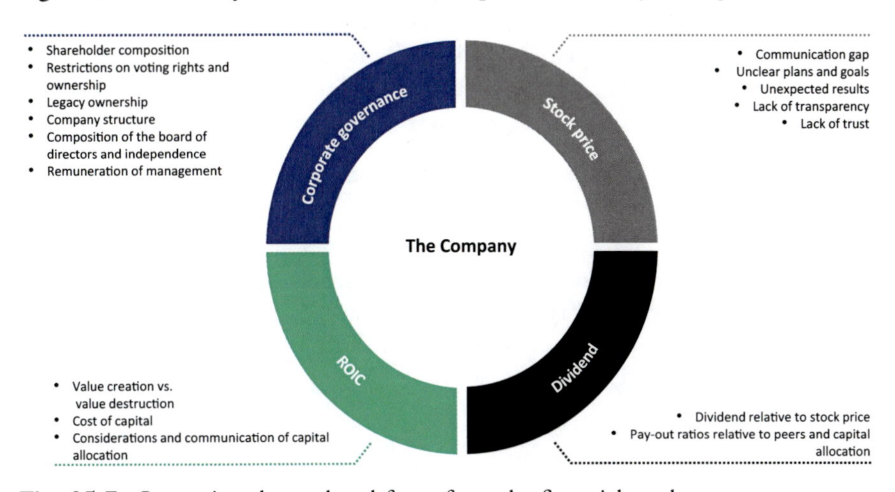

Fig. 25.7 Increasing demand and focus from the financial markets

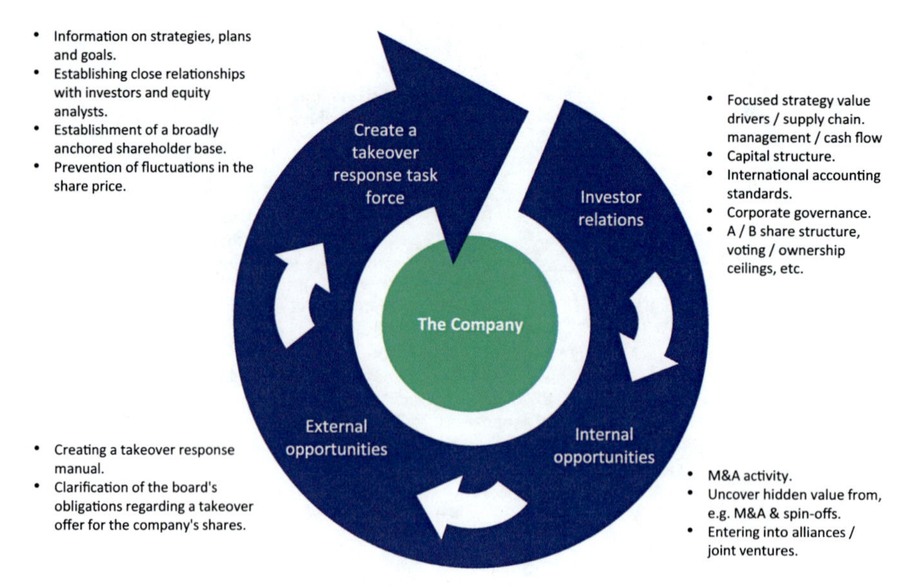

Fig. 25.8 Options for takeover protection

Particularly in listed companies with minority shareholders with no influence on day-to-day operations, the board of directors and the senior management are obliged to optimise the short-term and long-term return on the capital and the company they have been entrusted with managing. They also have a clear interest here in doing this. This is because it increases the likelihood that the same management will optimise its flexibility concerning the company's long-term strategic choices and its ability to make these choices on the best terms (Fig. 25.8).

Stock Exchange Handbook for Issuers

In the case of purchase offers, however all depending on both the relevant country of incorporation, and the relevant stock exchange on which the company is listed, there will typically exist rules for issuers (of shares) for an offer period as well as a number of other rules that both the offeror and the target have to follow. This will typically include that the target company's board of directors must publish a statement to the company's shareholders via the stock exchange before a certain period of time. In this statement, the target company's board of directors must assess the advantages and disadvantages of the takeover bid based on its knowledge of the company.

Such a statement must often be substantiated so that the company's shareholders can assess the background for the board of directors' recommendation, including matters the board of directors has emphasised. In this connection, the board of directors must look after the company's interests and thus all the shareholders, not only a group of them, and most certainly not their own personal interests.

There is often no requirement for the board of directors in its statement to make a specific recommendation to the shareholders to accept or not accept the offer, although this often happens in practice. However, the company can benefit significantly from having made some initial considerations about the argumentation that the company will use if the company suddenly receives a purchase offer.

Again, it needs to be reiterated that local legislators on takeover situations vary and that the selection of local advisors (corporate finance, lawyers, accountants, IR consultants, proxy adviser) is a must when preparing a proper and full TRM.

The Internal Improvement Opportunities

The internal opportunities for improvement include the areas in which the company itself has an influence, including:

- Focused strategy.
- Value drivers, supply chain management and cash flow.
- Capital structure, capital allocation, share buybacks and dividend payment.
- Application of international accounting standards.
- Best practice introduction and compliance with corporate governance recommendations.
- Elimination/modernisation of possible A/B share structure, voting and ownership limitations, etc.
- Incorporation of ESG as part of the company's strategy.

It is naturally recommended that the company focuses on the internal and external improvement opportunities, even before the company is suddenly in a situation where a takeover bid for its shares has been announced. If the company is well-prepared, and has obviously done its homework, before a bid for the company's shares has been submitted, the company's management will typically be praised by the financial markets.

If the company only takes initiatives after a takeover bid has been made or when the financial markets thinks that such could be on the way. In that case, management will often be accused of not having done its job well. This can be a problem for the management, cf. the description earlier in the chapter, if the company takes sudden initiatives in the form of, e.g. make jumbo dividend payments or make spontaneous acquisitions or divestments of Bus.

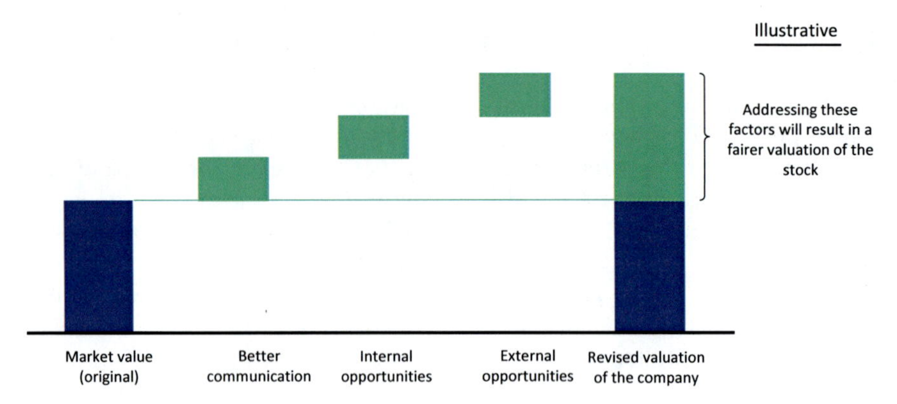

Fig. 25.9 Building blocks for closing the risk premium gap

The External Improvement Options

The company also has several external improvement options. These include several traditional but very significant initiatives. It can, e.g. be divestiture or spin-off into several companies, assuming that this is clearly in the interests of shareholders, which is often followed by acquisitions and alliances within the company's core area(s). We call these the external improvement options. However, they are very resource-intensive, require considerable effort and preparation on the company's part and often involve a not-insignificant risk (Fig. 25.9).

Not all companies reach all the way in the decision-making process regarding the various options for improvement measures that can uncover the company's hidden share price potential. However, it is a sound aspiration, and the financial markets will typically over time force the company in this direction (Fig. 25.10).

The measures mentioned above, all of which are described as shareholder-friendly measures, are recommended to be combined with preparing a TRM to fully safeguard the interests of the shareholders. Uncovering the hidden share price potential is the most future-proof hedge against a company's takeover (Fig. 25.11).

A Work Group List

A company should establish a full contact list, also referred to as a work group list. It should contain all contact and other practical information about the participants in the overall contingency and their possible role in the maintenance of the company's takeover contingency. As a rule, the list should consist of:

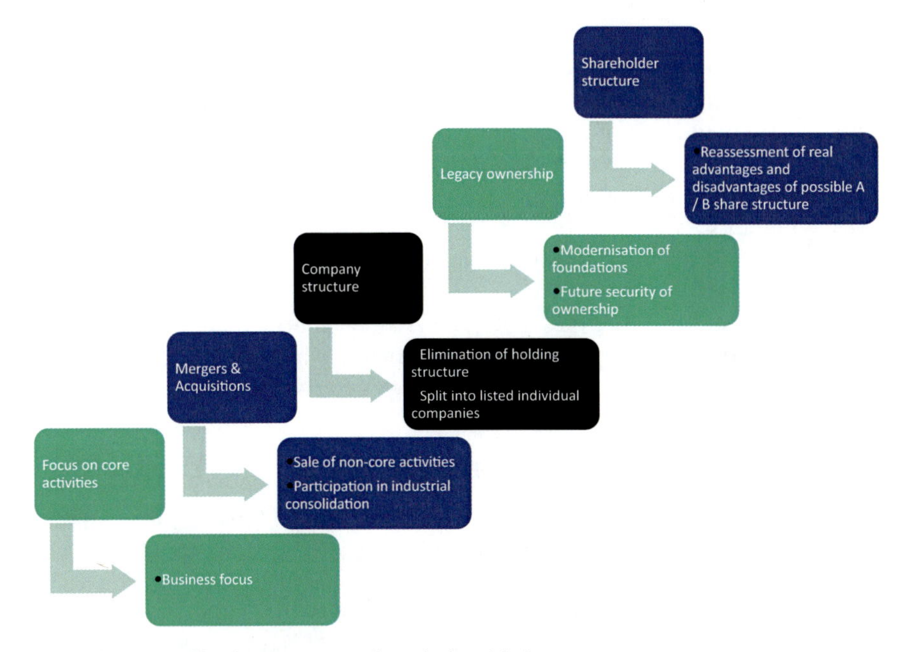

Fig. 25.10 Difficult issues are often dealt with last

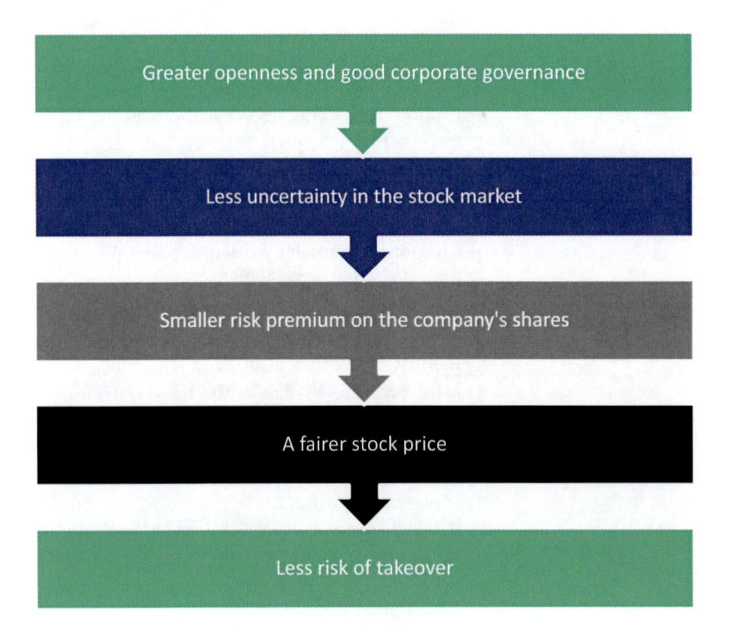

Fig. 25.11 Transparency reduces takeover risks

- The company's board of directors.
- The company's senior management.
- The company's key employees within IR, communication, legal, finance, accounting and any other critical company functions.
- Domestic and foreign corporate finance advisors.
- Domestic and foreign lawyer.
- Accountants.
- IR and financial communications advisor.
- Proxy adviser.
- The domestic and foreign equity analysts that follow the company.
- The essential domestic and foreign journalists who follow the company.
- Key people in the relevant stock exchange's Surveillance Department.

Keeping the Emergency Plan Up to Date

Emergency preparedness should always be up to date. However, it is important that the company only update what needs to be updated. Make it a habit to update contingencies every three months. If it is a fixed routine, each contingency update is less work to update.

The emergency plan	
Key message	• The version of history that the company will stick to • Maximum one key message and three sub-messages • It should be written so that the text can be used directly in a first press release
Fact and background	• Fact sheets with general and specific data about the area in which the crisis is located. It must be possible to be used as background material in an article. In addition, it must be written as short pieces of text on three lines saturated with facts
Verbal background information	• Talk sheet for the spokesperson who will brief the media about the story • It must be written in popular lay terms
Q&A	• Must provide answers to the most obvious questions, also the most obvious difficult and critical questions
Media training	• Have a standing agreement on media training which can be used when the company becomes aware that the crisis is imminent • It benefits credibility if the CEO stands up on behalf of the company and shows responsibility, in particular in the case of serious events
The company's profile	• The layman's version
Key data	• Also, data related to the crisis (extent of damage, etc.)
Contact person	• Make sure the contact is available and act as a filter between management and media, if necessary

Analysis of Shareholder Structure

In the case of a bid, it is essential to review the company's ownership quickly. Therefore, the company's TRM should also include a full analysis of the company's shareholder base, including:

- Listing of shareholders by size, geography and type.
- Summary of the largest shareholders' investment preferences and expected behaviour connected with a takeover bid for the company's shares.
- Contact persons for the largest shareholders at the top management and PM level.

Action Plan

The crisis arises	
N + 0 hours	The duty officer in communication is notified of the crisis and briefs the management and relevant persons on the prepared list
N + 1→ 3 hours	Press release sent out
N + 1→ 3 hours	Employees are briefed via an intranet, personal email or in meetings depending on how serious the crisis is (at the same time as above)
N + 1→ 3 hours	Relevant authorities are briefed on the notification. Arrange bilateral briefings (at the same time as above)
N + 1→ 5 hours	Relevant media briefs by telephone. Only from the time, the message is sent out. The IRO should simultaneously brief the company's most important shareholders and equity analysts on the situation
N + 1 day	One or a few journalists are invited to a briefing on the case with the CEO and a professional
N + 1 day	Spokespersons media trained
N + 1→ 2 days	Representative of the senior management meets selected journalists for a round table briefing
N + 1→ 3 days	The company briefs the public on the expected timetable for the details of the crisis, including when the case will be communicated next
N + x days	The company issues a press release stating that the crisis is over, how it has been resolved and how the company will ensure that a similar crisis does not happen again
N + x days	Management collects with the employees what the company has learned from the crisis and how it will impact the company in the future. It can be brought as an interview on internal TV, intraweb or magazine

Knowledge of Valuation Is Key During a Takeover

In the overall work with preparing a TRM, the company valuation is included with a significant weight. The purpose of preparing a valuation of the company's shares in a takeover scenario and as part of the company's TRM is to ensure that the board of directors gets a sense of the company's potential market value compared to the company's current share price. The company is thereby more quickly able to assess, in connection with a possible takeover bid for the company's shares, if the takeover bid appears to be attractive to the company's shareholders, or not.

Let us clearly state that during the preparation of a TRM, under no circumstances do we encourage to try to prevent or counteract a possible takeover bid for a company's shares—other than seeking to reduce the risk premium on the company's shares. That would be against good corporate governance, and in most cases current legislation. In connection with preparing a TRM, one of the first things to clarify is the obligations of the board of directors.

CHAPTER 26

Valuation in a Takeover Situation and Strategic Alternatives

The Use of External Advisers

The company is recommended to involve other external advisers (e.g. investments bank, lawyer, accountant and IR & financial public relations (PR) consultants, proxy adviser). For example, in connection with the discussed TRM, some elements may receive helpful attention from, e.g. an IR/communication advisor, corporate finance and legal. These situations include shareholder activism, M&A and other crises, where expert assistance is needed in the given situations.

Several parties are involved in preparing TRM. One of the resource-intensive phases in this process is the start-up phase, where the work must be organised, and the internal and external coordination must be determined.

The work should typically be structured in a steering group and several working groups. An investment bank's corporate finance usually works as project work that can coordinate with legal, senior management, and finance and directly with IR. Specifically for legal counsel, they assist a company in issues related to regulatory character or in case crucial external information liability is responsible.

The steering group is proposed to be led by, e.g. the company's CFO and with participants consisting of a representative from each working group, the IRO and corporate finance. Working groups should, for example, be established in the following areas:

- Law, accounting and tax, including lawyer and accountant.
- Valuation and financial markets-related matters.
- IR, communication, including an external financial communications consultant.

P. Lykkesfeldt and L. L. Kjaergaard, *Investor Relations and ESG Reporting in a Regulatory Perspective*, https://doi.org/10.1007/978-3-031-05800-4_26

- Proxy adviser.
- Administration and coordination.

We will not discuss further the details of a TRM as (a) the company's ambitions and resources vary; (b) local legislation and law vary from country to country, and between different stock exchanges; (c) much of the work are often driven by corporate finance who may each have a slightly varied approach.

The Preliminary Agreement with the Corporate Finance Advisor

The company is recommended to not only "play on one horse". An agreement with a corporate finance advisor is often made at the very last minute. Some corporate finance advisers' negotiation strategy may consist of waiting to bring the fee issue up until the company has no choice but to use the corporate finance adviser in question. This is because the corporate finance adviser at the time is so well versed in the matter, among other things, by having done a lot of preparatory work free of charge.

On the other hand, other corporate finance advisers, in a competitive situation, wish to get a signed mandate letter on the collaboration in place as soon as possible to know that other corporate finance advisers are not getting the task.

It should also be mentioned that it is in connection with takeover response tasks that the corporate finance adviser traditionally often receives the relative highest fees—precisely because it is in the situations where the company is most pressured.

Therefore, it is recommended that the company seeks several dialogues prior to the initiation of the work with a TRM. This means that two competing corporate finance advisers can present themselves as well as give an indication of the terms on which they would assist in a takeover response situation. In addition, this gives the company the advantage that it has tentatively negotiated terms in a situation where it is not under time pressure.

There are clear advantages of entering into an early preliminary agreement with a corporate finance adviser. In addition to an early fee discussion, the company also has the time to discuss with the corporate finance adviser in good time. This allows them to be up to speed on the case early.

It is not inconceivable that the company should pay the corporate finance adviser a nominal retainer (e.g. a monthly fixed fee) for the work of valuing the company as the corporate finance advisor cannot reliably assume that the collaboration will lead to a transaction. An offer to buy the company's shares should preferably not have been realised when the company establish a TRM.

Once the work is in process in terms of preparing a TRM, the company may continuously assess the quality and efforts of the corporate finance adviser in the same way as all other externals are continuously evaluated for their efforts and about the fee paid.

Valuation

Valuation (see Part 1) is one of the most important ingredients in the company's readiness for acquisition. Therefore, the corporate finance advisor adds several dimensions related to historical trades of comparable companies and BU. These additional dimensions consist of an analysis of:

- What market-related key figures have listed companies been traded for in the industry in question?
- What have bid premiums (i.e. the percentage premium to which comparable companies have been taken over about the share price that prevailed before the takeover bid) listed companies been traded at in the industry in question?
- What is market-related key figures comparable to unlisted companies traded for in the industry in question?

The corporate finance advisers prepare their analyses based on M&A databases, collecting and calculating transaction statistics based on completed corporate transactions. This analysis provides additional knowledge about the company's valuation of the DCF method. As with IR, valuation is considered an art rather than a science. Therefore, a company valuation requires considerable experience and a comprehensive understanding of the financial markets. As we have mentioned before, this is where corporate finance advisers have one of their absolute core competencies.

Therefore, every company must know that valuation is critical in a takeover situation. Moreover, critical decisions about the company's future are typically based on the corporate finance adviser's valuations. It is worth noting here that the corporate finance adviser may conflict with interest in certain situations, as its remuneration is generally transaction based.

A valuation can typically be divided into three levels, where variables such as resource input and purpose dictate which version is chosen for the given purpose (Fig. 26.1).

In connection with the preparation of a TRM, the company is recommended to select the preparation of a valuation, which is the middle of the three levels mentioned above. The valuation should include several value scenarios, including:

- The valuation that an equity analyst will typically arrive at.
- The valuation to which similar companies have typically been traded.
- A valuation by the company's high leverage, i.e. by high borrowing.
- A valuation in a break-up scenario, i.e. a sale of the company bit by bit.

The valuation result is, of course, confidential and for internal use only, and it is price-sensitive information. This is because if a valuation indicating a value lower than the current share price is published, it is likely to decrease its share

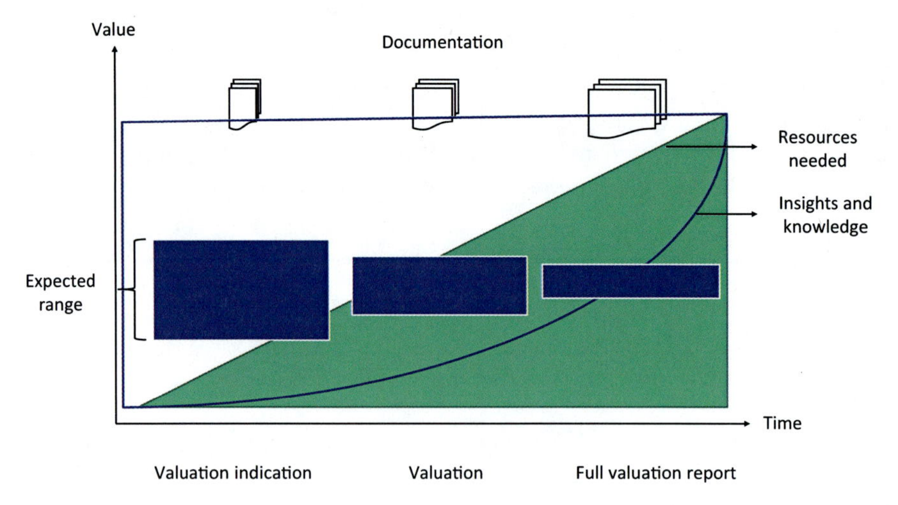

Fig. 26.1 Valuation requires resources and multiple layers of insights, knowledge and documentation

price. Conversely, a valuation above the current share price will increase the probability of a takeover bid for the company's shares. The company is not obliged to publish the result of such a valuation, partly because it is a valuation prepared based on subjective estimates.

Based on the valuation, the entire takeover response team, particularly the corporate finance adviser, can also better lay out a preliminary strategy for how the company can seek to ensure an even better price on the company's shares in connection with a possible takeover of the company.

Alternative Company and Ownership Structures

Further analysis must be made in connection with creating a TRM. The company's board of directors must examine the advantages and disadvantages of existing ownership structures, and possible A/B share structures, as well as restrictions on voting rights and ownership ceilings, which today exist in some countries.

We believe that an analysis of alternative corporate and ownership structures should always be conducted. Alternative structures should naturally always be in the interest of all shareholders.

Various studies have been prepared on which ownership best serves the shareholders, but it is difficult to conclude anything in general that proves sustainable in all contexts. Ultimately, it depends on the specific management of the company.

White Knights

In connection with the preparation of a TRM, a final analysis consists of identifying which peer companies, or strategic buyers, could constitute an appropriate white knight in a takeover situation. A white knight is a term for a company that a target company is looking for as an alternative buyer rather than the company, or the financial buyer, that has already given, or plans to give, a takeover bid on the company's shares. Therefore, this analysis will mainly be based on which peer companies:

- Has the best strategic fit in connection with a purchase of, or a merger with, the company, that is, can pay the highest synergy premium for the company's shares.
- Has the greatest synergy effects in connection with an acquisition of, or a merger with, the company, that is, can pay the highest strategic premium for the company's shares.
- Fits best in terms of culture and chemistry.

These white knights must subsequently be prioritised so that the company knows in which order the company may have to contact them in a takeover situation.

Management must first and foremost serve the interests of the shareholders in a takeover situation. This is done primarily by ensuring them as high a share price as possible. The advantage for the company of finding a white knight who is willing to pay the same or more than the party who launched the initial takeover bid, especially in connection with a takeover characterised by a bad mood between the parties, is that the management in the company is often able to negotiate better terms in the form of:

- More gentle demands for savings after a takeover.
- Greater security for the preservation of jobs.
- The board of directors and the senior management can, to a greater extent, maintain their position in the company or retire at reasonable terms.

Finally, it is also of great importance to the company that it has impacted its future ownership.

CHAPTER 27

Shareholder Engagement and Monitoring Market Activity

Corporate Governance

The board of directors must look after the interests of all of the shareholders of the company. The board of directors may thus not oppose an offer for the company's shares if this offer is in the interests of the shareholders and the company.

Whether this is complied with in practice in all cases can be discussed. It has been seen that some companies, e.g. companies controlled by a foundation, a family or another major shareholder, have prioritised interests that are not necessarily in line with those of minority shareholders, which usually prioritise the highest share price.

In some cases, this may lead to an intervention of the stock exchange or the local financial authorities. When the concept of good corporate governance is dealt with in this chapter, this may also be seen in the light of a TRM.

Recommendations for good corporate governance may in our view be divided a number of areas, e.g.:

- The interaction with the company's shareholders, stakeholders and other interest groups.
- Tasks and responsibilities of the board of directors.
- Composition of the board of directors.
- Remuneration of the board of directors and senior management.
- Risk management.

P. Lykkesfeldt and L. L. Kjaergaard, *Investor Relations and ESG Reporting in a Regulatory Perspective*, https://doi.org/10.1007/978-3-031-05800-4_27

The Role of IR

Internally, IR is a vital internal stakeholder like media, communication, legal and line managers. We have described that the IRO should strive to have strong relationships within the organisation (Chapter 19). It is noted that the IRO naturally has no business responsibility for the line managers and may therefore have a more lenient relationship with the line managers relative to the senior management. Therefore, IR has a clear role within IMC planning to align the internal stakeholders in crises (Fig. 27.1).

Through its work, IR is very close to the financial markets and is proposed to play a significant role in preparing the company's readiness for takeover. In addition, the company typically hires the services of a corporate finance, which can be significantly involved in the process. Still, this preparatory work should, to a significant extent, be driven by the company, not the corporate finance adviser, in order not to become too dependent on a particular corporate finance adviser at a time when a takeover is not relevant at all. In this connection, IR is proposed to have a leading role in close collaboration with the company's legal department (or external general counsel), and the CFO, who directly reports to the company's CEO, who in turn report to the board of directors.

For some companies, particularly SMEs, the IR function can largely be outsourced to confidential and trustworthy IR consultants. Their role is typically most significant in two situations:

- When critical situations arise or negative influence—be a discussion partner who may assist in the situation.
- Conducting perception studies to understand how external stakeholders anonymously and honestly judge the company.

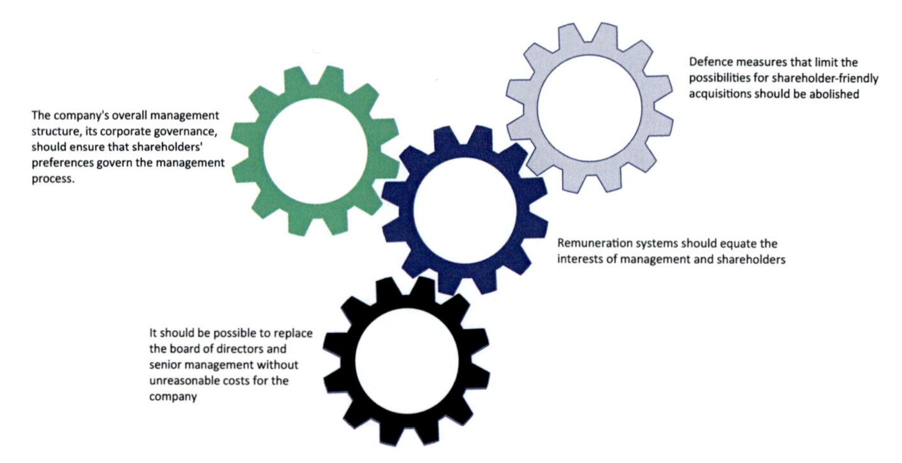

Fig. 27.1 Critical areas of good corporate governance

Monitoring Market Activity

The IRO monitors the market activity and ensures that shareholder engagement is responsibly constant and consistent. Monitoring market activity is a proactive method to follow the development of the company's shares irrespectively of the business model. The main areas of monitoring are:

- **Major shareholders:** Review a list of major shareholders in detail and know them. Valuable tools are investor perception studies, hereby identifying ongoing satisfaction and levels of trust.
- **Equity story:** Continue developing the company's equity story and communicate it effectively to the financial community; make sure guidance is both robust and credible and that shareholders have proper visibility concerning the company's value drivers.
- **Flow:** Track trading flows to decipher whether stake accumulation is taking place.
- **Activity:** Monitor trading activity relative to historical averages.
- **Voting rights :** Maintain an up-to-date view on the implications of voting rights and dual voting right structures.
- **Short positions:** Maintain awareness of short positions and related derivatives trading as indicators of the potential activist activity or a potential acquirer building a major shareholding.

Most of the above information is public, however time-consuming to compile. It is possible for the IRO to produce an activity report (applying the "digital IR tools" described in Chapter 21) or pay an external adviser to compile such a report. However, we suggest an alternative, or at least supplementary, method. It is part of the broker's and the equity analyst's job description to monitor market activity in the listed companies on which they have research coverage. Further, they have a clear interest and motivation to maintain a strong relationship with listed companies. Therefore, an IRO may utilise this position and collaborate with a broker to obtain market activity reports, while applying an arms-length principle.

CHAPTER 28

Investor Activism

It is not the scope of this book to perform a detailed scientific study of investor activism. Our approach is a pragmatic one, where we seek to provide value to companies' board of directors and senior management through an understanding of investor activism, including preparing for it, avoiding it and if necessary, dealing with it.

Our thoughts are universal and may in selected situations be more in tune with certain geographical continents than others. However, the financial markets are getting more global by the day, assisted by relevant regulatory framework from one continent being fully or partly adopted by other geographical continents.

As to preparations, actions and contingencies recommended to companies, some elements are similar to our listed recommendations set out under preparation of takeover response manuals, e.g. the company's and IRO's role, shareholder engagement and the use of external advisers.

A Historic Perspective

Investor activism has been known in one way or another for more than 60 years. In the 1980s and the beginning of the 1990s, corporate raiders frequently played a dominating role, often with a more short-term perspective to investor activism, and not always bringing good things with it other than short-term profits for the investor activists and other shareholders. Investor activism through those decades, particularly in the US, was often characterised by aggression, tough confrontation and winners/losers. Boards and senior management were typically, as it also often the case today, confronted with feared "white papers", i.e. documents or presentations running into the

P. Lykkesfeldt and L. L. Kjaergaard, *Investor Relations and ESG Reporting in a Regulatory Perspective*, https://doi.org/10.1007/978-3-031-05800-4_28

hundreds of pages containing the company's narrative as the investor activist wished to portray it. These "white papers" were then often (in full or in part) shared with, directly or indirectly, the media to put extra pressure on the company.

The past ten years have witnessed a further general rise in hedge fund investor activism, to a large extent fuelled by easy access to capital and lower interest rates. Investor activism has historically thrived the most in countries such as the US, Canada and the UK, rather than in the EU, although Germany has had its fair share of investor activism in recent years. This is partly for cultural reasons and possibly also due to the two-tier corporate governance systems with a separate board of directors (also referred to as supervisory boards) and operational executive management (or senior management) often dominating in Europe, which arguably has led to a relative lead in terms of good corporate governance. Finally, voting restrictions and split-voting structures (A/B shares) have also played a role in selected EU countries.

Also, historically, mainly for cultural reasons, public investor activism has not been well-regarded in Asia. However, the increasingly global markets have changed this, combined with a shift in corporate culture. Finally, Australia has become a high-risk country in terms of investor activism during the latest decade.

Historically, investor activism has been sparked by cash overloaded and conservative balance sheets, recent fundamentally underperforming or capital-intensive businesses, too complex operations due to conglomerate structures, low dividend pay-out ratios, a general lack of shareholder engagement and incompetent management or board of directors. Typically, in a historical perspective, the focus of investor activists has been on a financial nature.

However, rather than dwelling on history, let us focus on the past around three years and future trends.

Investor Activism Today—And Future Trends

The number of more public investor activist campaigns focusing on traditional financial, governance or compensation issues is believed to have seen a slight decline over the past couple of years.

The reason for this is believed to be:

- Institutional investors have become more outspoken than ever in their dialogue with companies and are doing some work that investor activists would otherwise do. However, institutional investors are doing the job much less confrontational and in a less visual/public manner. Typically, via a close and private dialogue with the chairperson and/or the CEO/CFO. Typically, institutional investors and the company have the same aligned interests in the longer term. Therefore, unless the objective is to get rid of the chairperson or the CEO, matters are in our view often better solved via a constructive dialogue, e.g. regarding significant shareholder proposals, strategic initiatives, corporate action, board and management compensation, etc. It sometimes takes a little longer but may still be preferable compared to the resources a public battle may demand.

- Companies' IR activities and shareholder engagement approach have been professionalised. As a result, the best companies proactively reach out to all stakeholders, including their major shareholders and investors, to understand their views and interests.
- Many, in particular, larger companies have over the past decade worked in a dedicated manner (in some cases sparked by an outreach by institutional investors) on improving good corporate governance. Through this process, independent board members may have been elected, compensation committees may have been created, and the board of directors has become much more in tune with the interests of the company's shareholders.
- Investors have become increasingly influential. That has led to increased pressure on companies, which has led to the leading companies being more trimmed and probably being in a better operational and competitive shape than ever, in our view (this means naturally that other companies will be outcompeted and die, merged or be taken over).
- Today, many investor activists have a longer-term investment horizon than earlier, where a break-up of businesses and a subsequent swift sale often was the endgame.
- However, it is uncertain which effect COVID-19 may have had.

Poorly run companies should still look over their shoulders for investor activists. However, other than primarily compensation, we are likely to see a surge in ESG-related activism, e.g. workforce diversity proposals, environmental impact proposals, workforce remuneration issues, etc. Moreover, the ESG-related investor activism is expected to grow exponentially as investors get their ESG-related competencies upgraded and new ESG regulation is implemented.

When divided into market capitalisation and company size, we expect that investor activism may increase relatively among SME companies and decrease relatively among large-cap companies. The reason is that many of the larger companies will themselves perform housekeeping exercises regarding good corporate governance and ESG-related matters prior to patience running out with institutional investors or potential investor activists. A number of SMEs do not fully have the initiative and/or competence to press on for the necessary developments and changes in good corporate governance and the ESG-related areas.

Investor activism will most definitely continue to exist—and for good reasons, we believe, particularly when exercised constructively.

How Should Companies Look upon Investor Activism and Deal with It?

In our view, investor activism is, like other issues or events, something which may happen sooner or later. Hence, the well-run company must have a contingency (also referred to as manual or playbook)—like a takeover response contingency or other kinds of issues management contingencies.

Companies are recommended to ask themselves:

- Is investor activism good or bad?
- How can we avoid investor activism?
- If we cannot avoid investor activism, what do we do if an investor activist suddenly knocks on our door, possible via the media?

Is Investor Activism Good or Bad?

In our view, investor activism is good for the company and its shareholders if exercised appropriately, well, fair and reasonable. Just as we believe it is the case with active ownership exercised in an appropriate manner (i.e. typically more longer-term, larger shareholders). Appropriate investor activism is, we believe when institutional investors are in a constructive dialogue with the company, be that the chairperson or the CEO, depending on the nature of the matter, and where the investor is reasonable, well-prepared and has good arguments.

The amount of these dialogues has increased significantly over the past one to two decades as companies' boards have been professionalised due to the positive development in good corporate governance.

However, there may be situations with incompetent and unresponsive boards of directors and senior management; badly run companies; and with either no dominating shareholders, or a larger passive shareholder unwilling to exercise influence; where the investor activist needs to take more harsh measures in use other than a private dialogue with the company. Here, the investor activist may well team up with other investor activists which may ultimately lead to an unsolicited takeover attempt of the company.

How Does a Company Avoid Investor Activism?

It is not rocket science. If the board of directors and senior management consistently:

- Run the company effectively in the long-term interest of shareholders and other stakeholders.
- Clearly articulate the company's equity story and value proportion.
- Disclose not only good news but also bad news.
- Perform efficient cash management and capital allocation, and the company has an optimal capital structure.

- Run the company in an operationally optimal manner.
- Ensure transparent and segmented financial reporting.
- Are on the forefront of non-financial/ESG reporting.
- Uncover the hidden value gaps.
- Evaluate and pressure-test strategic and transactional alternatives regularly.
- Benchmark the company against industry peers.
- Review regularly governance policies and board composition and thrive towards best practice.
- Identify the issues that may attract investor activists' attention, and deal with them.
- Communicate proactively in connection with significant events, including M&A.
- Act proactively rather than reactively in the dialogue with investors and equity analysts.
- Ensure regular shareholder engagement and acknowledge that some of the contacts to the most important shareholders, which in some countries earlier solely rested with the CEO and CFO, today more appropriately takes place with the chairperson. In a potentially aggressive investor activist situation, a close early and historical relationship with the companies' major shareholders is key to winning their support rather than this support landing with the investor activist.
- If a possibility in respect to selective disclosure rules, consult the company's major shareholders before putting forward important shareholder proposals at the AGM or an EGM
- Cater retail investors through digital IR as they are often lack loyalty in investor activist situations
- Perform perception studies among equity analysts and institutional investors on an annual or biannual basis, executed in an anonymised manner by an external IR adviser (and sometimes mandated by the chairperson, rather than the CEO, CFO or IRO, to eliminate any possible conflicts of interest, cf. Chapter 21).
- Ensure that the IR function monitors market activity, including strange movements and trading patterns, and keep an updated shareholder register. If the company identifies an investor building a significant shareholding, do not wait until the potential investor activist flags a 5% shareholding. Instead, reach out early for a dialogue.
- Recruit and maintain a best-practice IRO, and seek a dynamic and proactive IR strategy
- If approached by an investor activist, including a hedge fund activist, then embrace and listen to them. They are typically very knowledgeable, very well-prepared and very worthwhile listening to, and the company can typically learn a lot from them. It just requires confidence and an open mind.

Well, if doing the above, then the risk of investor activism is approaching zero. So, just do it! ... one may say!

The above initiatives may come from an experienced chairperson, the CEO or CFO, or the best-practice IRO. Still, unless the chairperson fully supports the initiatives and the chairperson fully appreciates the merits of these activities, then it is a lost course from the beginning. The board of directors simply needs to be on top of issues related to IR, in particular, know and have knowledge of the company's important major shareholders. The experienced best-practice IRO will typically be the initiator and conductor of a number of the above steps and activities.

To avoid investor activists, it is key for the company not only to think like an activist investor—but also to think and act like a best-practice board that focuses on the best long-term interests of the company.

In conclusion, best-practice IR is key to avoiding investor activism and the best-practice IRO must also work closely with senior management and the board of directors.

If the Company Cannot Avoid Investor Activism, What Does It Do if an Investor Activist Suddenly Comes Out of Nowhere and Knocks on the Door—Via the Media?

Dealing with potential investor activism is, to some extent, like dealing with several other critical issues, e.g. preparing a takeover response manual or preparing an issues management contingency relevant for the company.

In such situations, the company is at the contingency preparation stage recommended to:

- Map issues based on, e.g. feedback from shareholder engagement and the results of perception studies among equity analysts and institutional investors.
- Ensure a constructive response, and possible action points, to any weak points.
- Engage external advisers, e.g. financial adviser, lawyer, accountant, financial PR/IR adviser and proxy adviser as with the preparation of a takeover response manual (or playbook), where one scenario may potentially involve addressing a public media campaign by an investor activist via all available channels, e.g. the regular channels of company announcements, the media, social media, institutional investor engagement, virtual meetings with retail investors, etc., and pre-prepare relevant draft announcements, etc.
- Ensure that the company has the right digital and social media (e.g. Twitter and LinkedIn) communications competencies, internally or via external advisers. Social media platforms are a powerful platform to counteract a potential aggressive investor activist campaign. However, an

aggressive campaign is seldom in the company's interest as the risks of a damaged reputation is significant.
- Performing fire drill exercises with all relevant internal and external parties if deemed relevant. The company must rehearse and be ready to communicate its story and narrative flawlessly and convincingly. Speed is of the essence and getting the company's narrative successfully out before the investor activist is a key to success.

As to the companies' recommended preparation, action and contingency, some elements are partly similar to our recommendations set out under preparation takeover response manuals, e.g. the company's and IRO's role, shareholder engagement and the use of external advisers. As with many other things in life, the better prepared, the better the end-result.

We will not in this book discuss specific strategies or tactics related to investor activism as such strategies and tactics vary from case to case depending on the concrete situation and the company.

CHAPTER 29

The Company—Before, During and After an IPO

In this book, we discuss a number of the situations where a company has interaction with the corporate finance departments of investment. On this background, we will also briefly touch upon the process in connection with an IPO. The below review intentionally takes place at a very consolidated level with a view to point at just a few issues of expected interest of a company which is about a process without any prior experience or considers an IPO. For at further deep dive we refer to literature which intention is to explain the IPO process in detail.

Before and During an IPO

The reason with a company may typically be referred to one of several of the below reasons:

- The need for new capital to realise the company's (well-documented) growth plans (this is the reason for an IPO, which investors always prefer more than anything).
- The need to make the shares in the company liquid due to change of generation, or employees or early angel investors wishing for a long-term exit).
- One or several PE funds or venture capitalists seeing an exit following the expiry following the expiry of expected lock-up agreements of anticipated (this is the reason for an IPO, which investors are never really pleased with).

P. Lykkesfeldt and L. L. Kjaergaard, *Investor Relations and ESG Reporting in a Regulatory Perspective*, https://doi.org/10.1007/978-3-031-05800-4_29

Naturally, if one or several in the company's board of directors or senior management has prior IPO experience, that is naturally a significant advantage for the company. If this is not the case, the company may decide to engage an independent IPO adviser, who is not remunerated via a transaction fee and may therefore act as a completely independent adviser to the company. This may be a financial adviser (or another adviser with substantial insight into the IPO process), who is not involved in, e.g. placing new shares, raising new capital, etc., but solely advise on selecting the different adviser, organising the overall IPO-related work and challenging the various adviser during the IPO process.

The IPO process, once decided by the board of directors, typically starts with either the appointment of an IPO adviser, or a financial adviser—both following a structured beauty contest. The total IPO team does not typically differ from the team related to a public takeover situation and includes:

- Investment bank/financial adviser (which includes the advisory/banking side, ECM and equity research—however, involved at different times and manners, all heavily regulated by laws, regulations and compliance departments).
- Lawyer.
- Accountants.
- IR and financial PR consultants.
- Proxy adviser.
- Potentially different business-oriented advisers or consultants.

The total duration of an IPO process often takes up to twelve months, although it may be done in less than six months if the company is very well-prepared.

Without going into further depth, we illustrate below a typical IPO process although it is emphasised that the overview is simplified, summarised, by no means exhaustive, and may vary between different IPO transactions depending on the characteristics of both the company and the transaction. Such a transaction overview will typically be crafted and maintained by the financial adviser, where all financial advisers have their own way of illustrating the IPO process (Fig. 29.1).

During the IPO process—other than a long series of legal, financial and accounts-related tracks—the company will typically focus on:

- Getting a prospectus prepared with the assistance of its advisers—where in particular the company's equity story is completely essential. This is the one that investors buy into if they later decide to participate in the IPO.
- Getting the company's website updated, including preparing a detailed IR site.

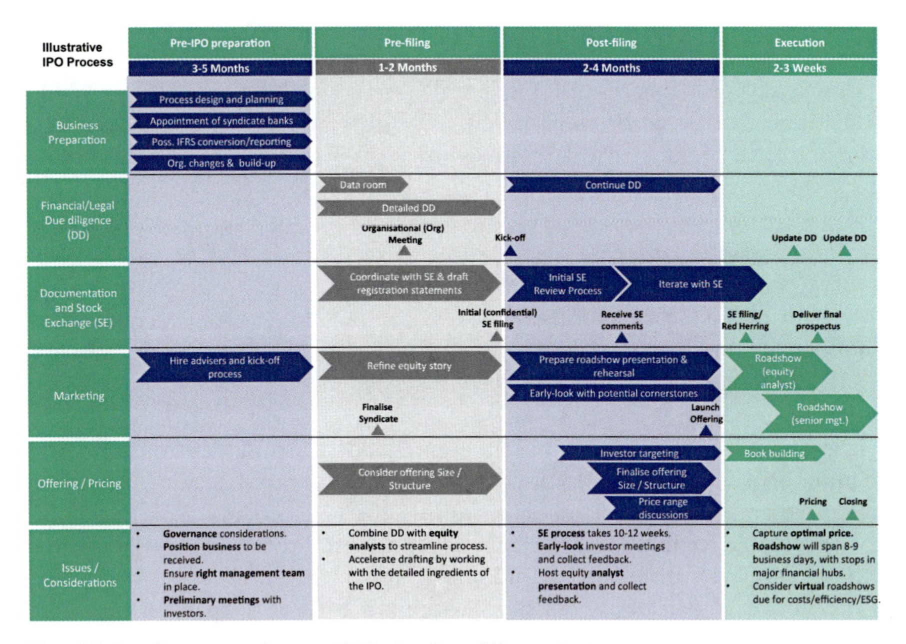

Fig. 29.1 Corporate finance IPO timeline (illustrative)

- Getting senior management rehearsed to for investor presentations, roadshow, dealing with investors and equity analysts—as well as the media.
- Ensuring that the board of directors and senior management really know of the future demands of being a listed company.
- Ensuring that the company—already before it becomes listed—has an IR policy, strategy and plan, and the knowledge and infrastructure to deal with investors, equity analysts and the media following an IPO, including creating an IR function and possibly recruiting internally or externally the coming responsible of IR (unless the company is a SME company where the CFO or CEO will deal with the ongoing IR work).
- With its IR and financial PR adviser, and entirely coordinated with the financial adviser, facilitates relevant pre-IPO communication, e.g. via the media, etc. However, this varies from IPO to IPO and ultimately, the financial adviser—who ultimately has the responsibility for the IPO and raising new capital—makes the final call on any financial PR-related activities.

After an IPO

A popular saying is that the IPO first really starts once the company is listed. Then, the company needs to focus on serving its new group of stakeholders and practice all of the activities discussed by us in earlier chapters of this book. This includes, e.g.:

- Following the company's IR plan.
- Be focus on news flow and follow-up.
- Focus on investor targeting.
- Focus on increasing analyst coverage and onboarding new equity analyst.
- And a long series of other activities.

It is important that senior management builds a strong relationship with the initial shareholders. In addition, the IRO should keep a detailed list of the onboarded and potential shareholders during the IPO process. Unfortunately, a lot of valuable feedback and findings are often lost if the IRO is not included in the IPO process from the beginning.

The above constitutes a holistic "a dipping the toe" regarding an IPO process, which a high proportion of company executives who have been through it characterises as one of the most exhausting business events they have been involved in.

A summary of facts and best practices is illustrated on page 329.

PART VI

Embracing Non-financial/ESG Reporting

CHAPTER 30

The Origin of ESG

> Without a sense of purpose, no company, either public or private, can achieve its full potential. It will ultimately lose the license to operate from key stakeholders.
>
> —Larry Fink, CEO of BlackRock

Social Awareness and Social Justice Dates Back Years

In the thirteenth century, the double-entry bookkeeping system was invented in Italy.[1] The method continues to be the dominant tool for companies and organisations to maintain financial information in a two-sided accounting entry system. Essentially, if all financial information affects at least two accounts by being both a credit and a debit, they must equal the same and therefore drastically reduce the margin of error and allows easier error detection in financial accounts. The "single bottom line" has been coined as it equals the same. In a sense, a well-aligned term to the neoclassical economic purpose of a company balancing revenue and costs to generate a financial profit.

Moving fast forward to today, breakthroughs in accounting have been limited. Instead, the focus has continued to be measuring a company's success and failures in terms of its overall financial P&L. Despite the century-old Greek term "philanthropy", there are only accepted principles of accounting. Still, a comparable social impact standard or other societal benefiting attributes remain non-existing.

Society has witnessed a fuelled focus on social awareness, social justice, environmental impact and economic marginalisation in the past decades. Likewise, investors, equity analysts and senior managers have demanded and attempted to measure a company's ESG or non-financial profit. This includes

P. Lykkesfeldt and L. L. Kjaergaard, *Investor Relations and ESG Reporting in a Regulatory Perspective*, https://doi.org/10.1007/978-3-031-05800-4_30

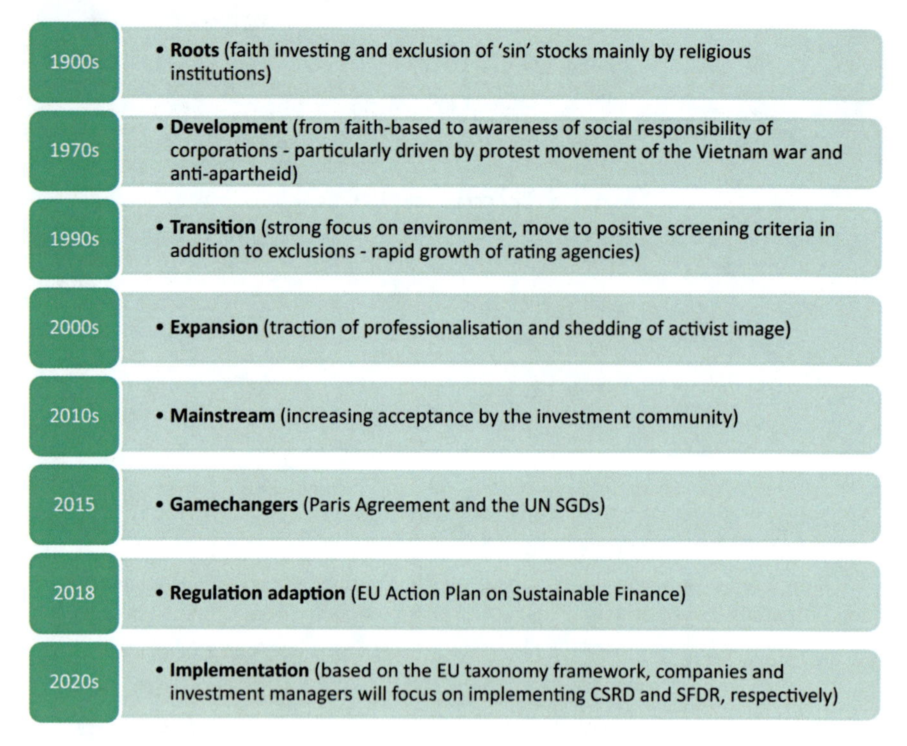

Fig. 30.1 Seven phases of ESG development (*Source* Loche et al. [2012] and McTavish [2018])

the company's exposure and effects factors, including environmental, social and corporate governance factors. These factors essentially consider the total cost of doing business and the greater good that a company provides.

Stages of ESG Development

In this book, we not deep dive into a literature review on the history and evolution of corporate social responsibility, as the reader can find comfort from other sources.[2] Considering both ethical and sustainable investing, the oldest method within the spectrum is the exclusion method. However, the scope of sustainable investing has increased in maturity and developed steadily in the past century. Yet, in the past two decades, the financial community has witnessed a "ketchup effect" in investing following the 2008 financial crisis, social and justice, environmental impact and economic marginalisation.

In 2012, three researchers pioneered and identified the five phases of responsible investments, and in 2018, the Norwegian asset manager, DnB, expanded on the universe and addressed an additional two phases[3,4] (Fig. 30.1).

We argue that the phases started in the 1900s, with religious organisations focusing on investing based on their faith and excluding "sin shares". The field became more mainstream in the 1970s/1980s, moving from faith-based activity to public awareness and social responsibility and justice—particularly in the light of the anti-Vietnam war and the anti-apartheid movements across the US and Europe. Environmentalism began to be recognised as coined by Rachel Carson.[5]

In the 1990s, the agenda of climate change gained movement, also resulting in the introduction of renewable energy-based investment funds and the term "sustainable development" coined by Gro Brundtland in 1992.[6] During the 2000s, the field of ESG expanded rapidly, with coherent ESG investment strategies being introduced.

In the 2010s, responsible investment and ESG became mainstream in the financial and regulatory community. The 2020s will likely be remembered for implementing the regulatory framework and harmonising ESG data. Like the rise of compliance costs following the 2008 financial crisis, we can assume that there will be an increased hunt for ESG talent that can assist investment funds and companies to improve their ESG approach.

Major institutional investors and regulators have embraced ESG development with the framework from the early pioneers. As a result, we see a highly established field that all companies must seriously consider. To illustrate the phases of ESG development for the past decade, we have outlined the communication development of the largest asset manager in the world, BlackRock, and the world's largest investment fund, Norway's sovereign wealth fund, see also below.

A Brief History of Non-financial Reporting Frameworks

Therefore, there have been ideas to better register and track a company's social performance. Still, the majority have failed to find a transparent, tangible and numerical method to track non-financial profit on par with accepted financial profit measures. The most notable measures[7] include double-bottom-line consideration (1984) (considering "social impact" alongside "financial profit"), the previously mentioned Elkington's triple-bottom-line framework (1994) (environmental, social and governance) and the quadruple bottom lines (2002) (including "cultural" or "spirituality").

In addition, several alternative frameworks of the spectrum include the social return on investment (2000) (SROI), the environmental profit and loss approach (2010), and Boston Consulting Group's total societal impact framework (2017). Lastly, the financial literature has published a wide range of papers on impact investments, capital models, full cost accounting and information advantage anomalies through non-financial factors throughout the past decades.

The critical aspect is that all the above measures are the lag in general and wide acceptance—like the recognised principles of the double-entry booking keeping system. However, in the UN's Principles of Responsible Investment paper of 2006, the organisation "blue stamped" ESG as an umbrella term to enhance corporate social responsibility encompassing various frameworks. With an accepted term and definition, institutions could then subsequently translate into transparent and accepted non-financial measures for companies to report to establish and track their general good for society.

References and Sources

1. Lee, G. A. (1977). The coming of age of double entry: The giovanni farolfi ledger of 1299–1300. *Accounting Historians Journal* 4(2): 79–95.
2. Agudelo. M., et. al. (2019). A literature review of the history and evolution of corporate social responsibility. *International Journal of Corporate Social Responsibility.*
3. Loché, C. et. al. (2012). From preaching to investing attitudes of religious organisations towards responsible investment. *Journal of Business Ethics* 110(3).
4. McTavish, L. (2018). ESG: being a responsible investor is no longer limited to excluding stocks. *DNB,* https://dnbam.com/se/finance-blog/esg-we-have-come-a-long-way-since-the-first-exclusions-of-sin-stocks.
5. Carlson, R. (1962). Silent spring. Hougton Mifflin.
6. Brundtland, G. H. (1992). Brundtland Commission report: Our common future. *UN Rio de Janeiro summit.*
7. Douglas, B., et. al. (2017). Total societal impact: A new lens for strategy. The Boston Consulting Group.

CHAPTER 31

Stakeholder Capitalism and Sustainable Leadership

Including Stakeholders in Neoclassical Economics

Global investors were already embracing ESG, and this is expected to accelerate because of SFDR and the amendments to MiFID II. Institutional investors will be obligated to report on non-financial data and considerations. To attract a sustainable inflow of capital in funds, the financial community will assist other stakeholders in mobilising non-financial reporting. The more transparency on the topic, the more (negative) focus will be on companies with inadequate non-financial disclosure practises and their approach to sustainable leadership. If a company has a limited focus on non-financial reporting, we will argue that, especially in the light of SFDR, such a company will eventually dry out its access to new capital from the financial markets.

The COVID-19 pandemic's impact has been a critical linchpin for stakeholders to increase pressure on companies and senior management to broaden their focus from only financial profit. COVID-19 has insofar led to megatrends such as:

1. Fundamental changes in consumer purchasing behaviour and focus on long-term sustainable development.
2. "The great resignation" disruption of the job market forcing companies to rethink methods of retaining talent.
3. Regulator's method of tying stimulus packages with green initiatives via the EU's green deal and the US' "GreenNewDeal".
4. Increased focus on social distance and health considerations to limit disease outbreaks and thus protect the vulnerable citizens.
5. An accelerated focus from internal and external stakeholders on identifying, managing and considering a company's non-financial performance.

P. Lykkesfeldt and L. L. Kjaergaard, *Investor Relations and ESG Reporting in a Regulatory Perspective*, https://doi.org/10.1007/978-3-031-05800-4_31

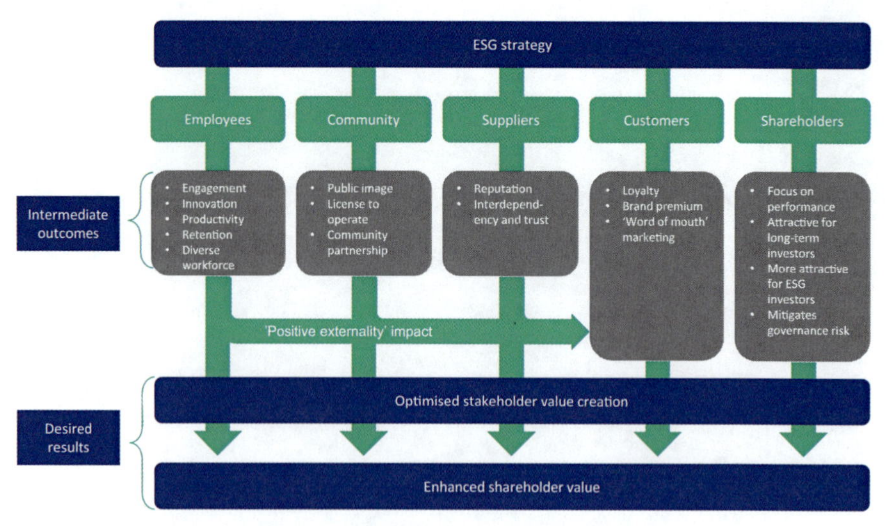

Fig. 31.1 Stakeholder model and ESG (*Source* Kay et al. 2020)[1]

Therefore, there is a clear need for a framework for stakeholders to track the non-financial performance and status of a company to ensure that they comply with their promises—and allow stakeholders to compare a company's performance relative to its peers. Much like financial reporting, this framework can also assist in setting remuneration (and bonus) structures, long-term strategies, product design and share CAPEX and M&A plans.

Closely aligned with Freeman's stakeholder model, the US organisation "Business Roundtable" (BRT) published the stakeholder model of ESG in 2018, supported by over 180 CEOs of market-leading companies. The motivation was a clear aligned framework to move from a shareholder-driven framework to considering and driving ESG on all stakeholder levels. So, despite assessing decisions in terms of the intermediate outcomes and impact from "externalities", business leaders will include ESG strategy on all decision-making levels to optimise shareholder and stakeholder value (Fig. 31.1).

An important consideration is to consider the externalities from an ESG perspective. Examples include lowering employment churn if employees' values are aligned with the company's strategy. Likewise, public image from the local community and reputation among suppliers have positive external impacts on the company's customers. Therefore, their goal is to optimise stakeholder value creation and enhance shareholder value.

Raising the Bar on Stakeholder Capitalism

Considering the UN's SDGs (Chapter 16), the UNGC launched the campaign "Making global goals local business" in 2016. The focus was a step-change to assist companies in transforming their focus to include core SDGs along with universal principles at the centre of its strategy, operation and approach to stakeholders. As a result, the UNGC has successfully created the world's largest sustainable initiative and has become an attractive platform for CEOs to exchange learnings within corporate sustainability.

The world is lagging behind the UN's SDGs.[2] Out of the 35 sub-categories in the 17 SDGs, only five goals are "on track", ten goals are classified as "progress needed", fourteen "no progress" is made, and six goals are "in deterioration". Therefore, pressure from the financial community and stakeholders will continue to increase to alleviate social, economic and environmental problems. In the UNGC's 2020 progress report, it is highlighted that 90% of businesses have integrated its objectives, and 85% have selected which SDGs to focus on.[3] However:

- Only 46% of companies have aligned their core business with their selected SDGs.
- Only 39% set goals that are sufficiently ambitious, science based and aligned with the SDGs.
- Only 45% of companies track their actions versus its SDGs.
- Only 31% asses their negative impacts on the SDGs.
- Only 13% of businesses act through their suppliers to influence their SDGs.

The UNGC has launched the "decade of action" to re-establish the step-change as a call for companies to increase their ambition levels by fully embedding its principles and the SDGs. This must be implemented top down from a strategic level, i.e. starting with the board of directors, and filtered through to the operational and stakeholder levels.

The organisation has outlined the "implementation model", a systematic and logical approach to increase a company's ambition level.[4] It is the board of directors' fiduciary duty to act on behalf of the investors to ensure that ESG is fully integrated into the business, and this includes driving ESG implementation from the top. The model commences in three levels: (1) anchoring ESG in purpose, governance and corporate business strategy, (2) cascading ESG across all functions to deepening operational integration and (3) communicating ESG as part of the business to enhance stakeholder engagement (Fig. 31.2).

Including SDGs in the business strategy and purpose shows a strong signal to the company's stakeholders that they are supporting the global transformation. It is increasingly expected that a company has fully integrated ESG into its strategy and that it also relates ESG to its global transformation. A company must highlight which SDGs the company affects (positive and negatively), how they may be improved, and report on the progress and goals in a transparent manner. This includes both production considerations (e.g. energy, water, emissions) and non-production considerations (e.g. diversity, education, well-being).

SDG IMPLEMENTATION FRAMEWORK		
ANCHORING AMBITION IN STRATEGY AND GOVERNANCE	**DEEPENING INTEGRATION ACROSS OPERATIONS**	**ENHANCING STAKEHOLDER ENGAGEMENT**
PURPOSE	**PRODUCTS AND SERVICES**	**REPORTING AND CORPORATE COMMUNICATIONS**
• In articles of association • In vision/mission statement • In company values	• In development and innovation processes • In supply chain management and across procurement • In Total Quality Management System	• To investors and shareholders in financial statements • In non-financial statements and third party reporting • In public relations and communications
GOVERNANCE	**PEOPLE MANAGMENT**	**SALES AND MARKETING**
• On board agenda and committees • In board competencies and diversity • In executive recruitment and incentives	• In performance management and remuneration • In training and learning • In company culture and communications	• In all brand and product promotion • In all customer engagements • In consumer education and behaviours
CORPORATE STRATEGY AND GOALS	**CORPORATE FINANCE**	**PARTNERSHIPS AND STAKEHOLDER RELATIONS**
• In ambitious goals and targets integrated into a balanced scorecard • In innovative business models • Integrated unit business strategies	• In key business investments • In allocation of capital and financial strategy • In managing risk and revenue	• In forming alliances to accelerate impact • In relationships with communities and stakeholders • In ensuring social license to operate

Fig. 31.2 SDGs implementation framework (*Source* UNGC 2020)

The Advantages of Non-financial Reporting

Maximising shareholder return has always been the primary goal of a company. However, companies are urged to consider all company stakeholders with an ongoing societal change. In addition, the new world of business and the financial markets are increasingly dictated by ESG and non-financial reporting. Therefore, embracing non-financial/ESG reporting must be an integral part of a company's strategy formulated at the board of directors level.

- Internal stakeholders of employees, managers, senior management, the board of directors and external shareholders are urging companies to have an active approach to ESG.
- External stakeholders, including the local and broad community, suppliers, customers and regulators, recognise that the UN's principles and goals cannot be met without the active involvement of companies.
- The inflow of capital from institutional and retail investors into ESG-compliant companies has been growing at an immense rate over the past decade.
- Methods, tools and platforms that include and track ESG-related data are increasing in importance, which will discredit the non-compliant companies.
- Several ESG attributes can positively impact a company's operations and thus its financial measures (Chapter 5).
- There is established and increasing evidence that companies with a strong ESG presence and transparent non-financial reporting culture

outperform their non-compliant peers. This evidence is unambiguous, as ESG-compliant companies tend to minimise their downside risk.

References and Sources

1. Kay, I., et al. (2020). *The stakeholder model and ESG*. Harvard Law School Forum on Corporate Governance.
2. Rasmussen, E. (2020). *The sustainable business pioneer: A portrait of Lise Kingo*. A Sustainia Publication.
3. UN. (2021). *Sustainable Development Goals Progress Chart 2021*. https://unstats.un.org/sdgs/report/2021/progress-chart-2021.pdf.
4. UNGC. (2020). *SGD ambition: Scaling business impact for the decade of action*. United Nations Global Compact Library.

CHAPTER 32

Institutional Investors Are Embracing ESG Strategies

Defining Responsible Investing

It is yet not clear how non-financial reporting can completely be integrated and all the practicalities surrounding ESG. However, responsible investing is more straightforward and can be separated into ethical investing and sustainable investing. The two methods are similar, as investment decisions are based on non-financial reporting factors. In a way, it judges a company more intensely on corporate governance and influences its non-financial reporting standards on an investment level.

Ethical investing arises from the investor's values, where the investor is willing to compromise expected returns with a good conscience. Typically, they exclude specific industries (e.g. alcohol, tobacco, the arms industry, gambling and adult entertainment) or companies (e.g. high-interest rate loan companies or perhaps animal-product producers) from the portfolio. So, even though the investment case on financial reported figures may be solid, they are excluded for non-financial reasons. Nevertheless, despite sacrificing return, this can be a valuable approach to attract capital from all like-minded investors, such as pension funds with high morals.

Sustainable investing is somewhat more nuanced as the investor applies ESG systematically to the portfolio selection process, as listed below. These strategies are typically combined with the six classical investment strategies (Fig. 32.1).

Aside from viewing the tangible benefits of non-financial reporting, we believe there is room in the market for risk–risk investments. Many listed companies have disproven or untested business models but cannot attract investments. Notably, many pharmaceutical and IT companies have a limited

P. Lykkesfeldt and L. L. Kjaergaard, *Investor Relations and ESG Reporting in a Regulatory Perspective*, https://doi.org/10.1007/978-3-031-05800-4_32

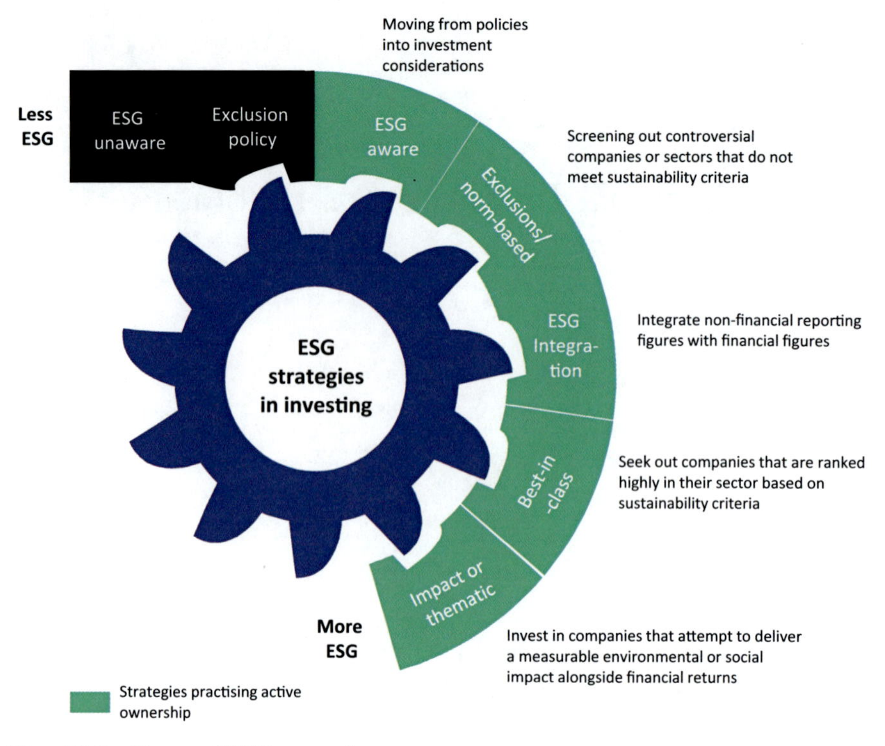

Fig. 32.1 Types of ESG strategies

proven track record, but investors are willing to pay a large price for the possibility of a successful investment case.

Likewise, we believe this will gradually overflow to the sustainability sector, where companies with a solid idea and business plan to assist in the SDGs will find easier funding opportunities. Increasingly, institutional investors and the overall financial community pressure companies to take ESG seriously. This includes pressuring companies to increase diversity of their boards, sustainable investing of the company's employees' pensions, transparent tax reporting, etc. Besides investing in a company, institutional investors are adapting an active ownership approach to pressure companies to be more ESG compliant similar to the company's other stakeholders. In addition, institutional investors are teaming up together to increase the pressure. Therefore, the faster a company may adopt to best-practice non-financial/ESG reporting, the better a long-term return to its shareholders in our view, all things being equal.

Positive Teachings from BlackRock

As a custodian of 9 trillion euro (10 trillion US dollar) in assets under management, BlackRock, Inc., serves as the largest asset manager in the world. In 2021, around 30% of the total involved ESG integrated active and advisory strategies across 1,200 sustainability matrices. Given BlackRock's substantial size and equity ownership, its CEO, Larry Fink, has published an annual letter to CEOs globally since 2012 (except for 2013 for unknown reasons). The topics have concentrated on different themes important to creating long-term value creation for a company's shareholders:

- In 2012, the letter stressed that CEOs should focus on good corporate governance practices resulting in improved long-term business performance. With focused best practices, the company should be transparent and honest towards their shareholders on corporate governance issues and include shareholder feedback in its approach to shareholder meetings.
- In 2014–2015, Fink disfavoured cutting short-term CAPEX to increase dividends and share buybacks. Balancing capital strategy is wise, however, not if it jeopardises a company's ability to generate long-term results. In 2015, the letter even stated that BlackRock revised its proxy voting guidelines to vote against directors, where there was evidence of short-termism afflicting ineffective corporate governance.
- In 2016, Fink urged CEOs to outline their strategic framework for long-term value creation to combat continued raising dividends and share-buyback pay-outs. As a result, the financial report of companies should be concerned with long-term considerations on navigating the competitive landscape, technological disruption, geopolitical events, relevant financial metrics and compensation schemes. As a result, quarterly financial guidance should also be limited, resulting in short-termism. In the letter, Fink also mentioned ESG for the first time. Frank and brief, Fink said they were important to have a quantifiable financial impact, and BlackRock plans to integrate them into the investment process.
- In 2017, considering the UK's Brexit vote, the former US President Trump's focus on domestic trade policies and tax reforms, and the growing back lack to globalisation and technological changes, Fink focused on globalism. Only mentioning ESG once, Fink urged CEOs to include macroeconomic themes in their strategic review and long-term considerations.
- In 2018, Frink focused on the term purpose, a topic we further explore in this chapter. Introducing a new model for corporate governance, a company should consider their overall purpose and approach to their stakeholders while questions like "what role do we play in the community? And how are we managing our impact on the environment?"
- The 2019, letter was titled "purpose & profit" and raised the subjects of social justice, public pressures, a new generation of investors and leadership. Essentially, Fink argued that purpose and profit are extricably linked and that is essential to consider when approaching all stakeholders and thus achieve long-term profitability.
- Adding climate risk as an investment risk in the letter of 2020, Fink saw sustainability integrated portfolios to provide a better risk-adjusted return

to investors. As a result, BlackRock made ESG funds standard building blocks for multi-asset solutions and more systematic active management solutions for ESG.

- Mentioning ESG four times, the 2021 letter concentrated on the scientifically established threshold targets for global warming. In addition, Fink focused on the newly developed TCFD framework and increasing transparency on ESG data and disclosure. Therefore, the central theme was not only to perform well on ESG matrices but also to report the status in a transparent manner.
- Almost removing the abbreviation "ESG" (only mentioned once), the 2022 letter is entitled "the power of capitalism". Fink affirms that fundamental capitalism is driven by mutually beneficial relationships, and therefore, including all stakeholders in a company's strategy and purpose is capitalism itself at its core. Sponsoring a Centre for Stakeholder Capitalism, Fink argues that companies that do not embrace their stakeholders will witness diminishing market opportunities.

The communication has continued to focus on long-term value creation but moved from themes of corporate governance, capital allocation, short-termism, global risks—to an ESG approach or stakeholder model to steadfast purpose and non-financial reporting. The newest theme in 2021 and 2022 is the transparency of non-financial reporting, also supported by the US' government accountability office (GAO) declaring in July 2020 that investors needed better ESG information, not more of it.[1] BlackRock advocates for the TCFD non-financial reporting (Chapter 35), and in addition to BlackRock, more and more investors and equity analysts are increasingly applying sustainability measures in their financial modelling.

Based the above, Larry Fink has personalised the institutional investors moved towards embracing ESG.

Positive Teachings from Norway's Sovereign Wealth Fund

With the mission to safeguard and build financial wealth for future Norwegian generations, Norway's sovereign wealth fund administers 1.2 trillion euro (1.4 trillion US dollar) and is thus the largest shareholder in the world. In addition, the fund serves as a financial tool for the Norwegian government savings—and has, across two decades, transformed from a classical diversified portfolio strategy to an active ownership model with a focus on responsible investing.

Today, the fund owns on average 1.5% of all global equity listed companies and includes a broad investment portfolio of ~9,000 companies in 73 countries. Since 1996, the fund has worked with a transparency and exclusion framework and therefore closely screens, monitors and invests in companies that fit the fund's policy standards. The company's website has a live-updated list of companies that it owns, the reason that they are held, ethical concerns and the fund's voting record on the company's AGMs.

The primary investment philosophy is to establish a standard ground filter through established principles across markets and raise the bar for all companies. Then exercise ownership by discharging obligations to vote at company AGMs, having ongoing investor dialogue and follow-up on business practices, opportunities and challenges. Finally, the fund implements an ESG framework to identify, measure and manage ESG-related risks that can impact the fund's financial and non-financial performance.

Being a large fund, reporting directly to the Norwegian financial Ministry and with an influential Council of Ethics established in 2004, the set-up has moved from a classical investment shareholder model to a stakeholder-driven model. Exclusion themes have in the past decades ranged from human rights, nuclear arms, child and forced labour, wiretapping, coal, fossil fuels and in 2021, requiring all companies in the portfolio to take climate change.[2]

For any IRO, it is crucial to recognise the high level of transparency that the fund has established. In addition, the exclusion themes and considerations of the fund are good early indicators of considerations the company should take from a strategic, communicative and IR perspective.

References and Sources

1. Fouche, G. (2021). Norway wealth fund calls on companies to act on climate. *Reuters: Sustainable Business*. https://www.reuters.com/business/sustainable-business/companies-should-take-more-action-climate-norways-sovereign-fund-says-2021-12-01/.
2. U.S. Government Accountability Office (GAO). (2020). *Public companies: Disclosure of environmental, social and governance factors and options to enhance them*. GAO. https://www.gao.gov/products/gao-20-530.

CHAPTER 33

Consideration Sustainable Finance Disclosure Regulation (SFDR)

Motivation of SFDR

The impact of the SFDR (and the associated amendments of MiFID II) on the buy-side industry and the synergies with a company's taxonomy reporting obligations are highly relevant to understand for the IRO. SFDR applies to all EU investment firms and alternative fund managers. It aims to protect end-investors of funds and limit the use of "greenwashing". A "comply or explain" approach necessitates all EU institutional investors (and investors marketing to EU investors) to make several sustainability-level disclosures at an entity and fund level (including funds and sub-funds). The disclosures are outlined in a framework of three concrete categories.

PMs, PE managers and other asset managers must have outlined and implemented ESG policies and considerations in their investment strategy. However, this is a time-consuming task given the early stages of ESG reporting and data harmonisation. In addition, due to the essentially non-existent ESG data on listed SME and unlisted companies (especially relevant for PE managers), the exercise remains almost impossible. As a result, the buy-side industry demands disclosure for the companies at an unprecedented scale; the best-practice should embrace this and collaborate with them.

The new amendments to MiFID II obligate investment firms that market towards EU investors to include their clients' ESG preferences in the suitability onboarding framework. Therefore, the investment firm must not only align their clients risk preferences to their specific portfolio exposure and investment advisory—but the investment firm must do the same for the clients' preferences (Fig. 33.1).

P. Lykkesfeldt and L. L. Kjaergaard, *Investor Relations and ESG Reporting in a Regulatory Perspective*, https://doi.org/10.1007/978-3-031-05800-4_33

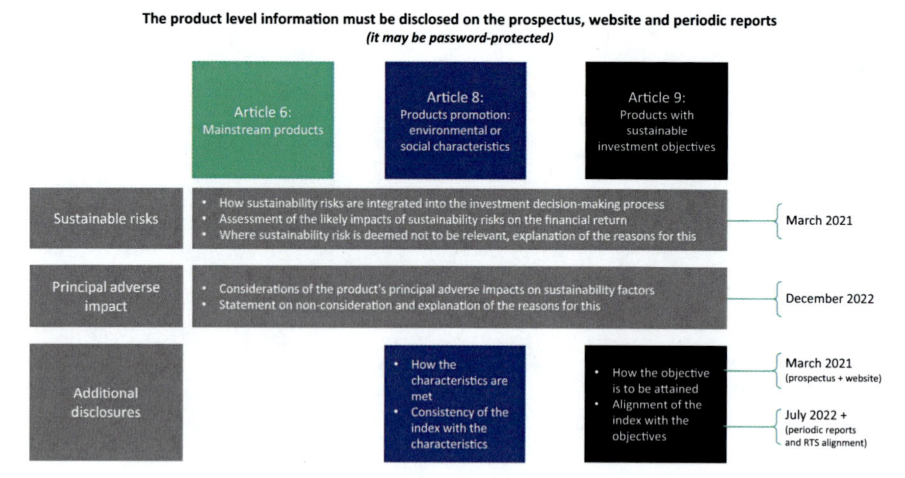

Fig. 33.1 Overview of SFDR (*Source* KPMG 2021)[1]

Disclosing ESG on Entity and Fund Level

At an entity level, an investment company must outline policies on the integration of ESG in the investment strategy. In addition, it must include disclosure if (how and why) it will implement the principal adverse impact factors (PAI) (PAI is: "Negative, material or likely to be material effects on sustainability factors that are caused, compounded by or directly linked to investment decisions and advice performed by the legal entity." Examples include GHG emissions, carbon footprint, unadjusted wage gap, board gender diversity, etc.) in its investment strategy—and provide corresponding remuneration policies alignment within the framework.

All investment companies with over 500 employees (including parent companies) are obligated to report PAI factors. Others are allowed to opt out but must then describe the reasoning behind it. If PAI is integrated, the company must publish, on their website, a detailed description of which PAI factors are considered and prioritised and its due diligence policies. As supported by the SFDR guidelines, the information may be uploaded on a password-protected client website—or provided to the investors in a physical format.

On an entity level, all investment firms (with over 500 employees) must have published an annual PAI statement (within the period of 10 March to 30 June 2021) considering its ESG progress and integration of PAI factors. Henceforth, this must be updated annually. Likewise, on a product level (described below), the investment firm must also have published a detailed disclosure report from 1 January 2022 ("starting of the reference period"). Future reportings include from 1 January 2023 (including a comparison to the reference period), then from 1 January 2024 (including a comparison to the reference period and previous year) and so on. In 2027, the investment company must include a five-year summary of its ESG/PAI progress.

From August 2022, the amendments of MiFID II require investment firms to include their clients' ESG preferences in the suitability onboarding. This includes asking and integrating their clients' wishes to a minimum proportion of sustainable investments defined as taxonomy activities (cf. fig. 16.6), SFDR fund classifications (including ESG subcategories, see chapter 44) as well as their clients' desire to integrate PAI factors. Additional requirements in terms of including ESG considerations in the manufacturing of new financial products are obligated from 22 November 2022.

Classification 1: Minimum Requirements (EU 2019/2088, Art. 6)

Article 6 contains the minimum requirements of SFDR and can be considered the sustainable risks. The article is applicable for all funds and requires their prospectus to contain either:

- Describe how the PM's investment strategy considers sustainability risks and impacts financial performance. This means how the PM deals with ESG risks and under what conditions an ESG event or new information can affect the investment's value, or:
- Describe which ESG risks (PAIs) are irrelevant for the investment strategy and the reasoning behind the determination.

The regulation mainly covers funds that do not integrate sustainability into the investment strategy and could include a company that ESG funds currently exclude. Examples include sin shares such as tobacco or coal producers. These companies can still operate, be listed and invested in, in the EU; however, the fund must disclose these investments and label them as non-sustainable. The implication is a fund which can face considerable marketing difficulties versus sustainable funds.

Classification 2: E&S Funds (EU 2019/2088, Art. 8)

Article 8, also known as "environmental and socially promoting", contrary to article 6, this article is concerned with sustainable funds that promote ESG characteristics. The article is relevant for funds that promote ESG characterising and requires them to provide specific details in the prospectus concerning:

- A detailed description of the fund's investment strategies concerning the environmental and social categories of ESG. The description must include which methodologies are established to assess, monitor and measure

them. "G" is less specified. However, the article assumes that good corporate governance is a prerequisite to the other characteristics.
- The fund must specify which sustainability indicators or benchmark index (not necessarily an "ESG-related") it follows and disclosure the consistency of the index relative to the fund's investment strategy.

Classification 3: Impact Funds (EU 2019/2088, Art. 9)

Lastly, article 9, also known as "funds targeting sustainable investments", covers funds targeting bespoke sustainable investments and applies. This is essentially the funds that use ESG as a fundamental objective and, unlike article 8, also use an ESG-orientated index as a reference (if they use an index). The article is relevant for funds that have a sustainable investment objective, and in the prospectus, the fund must:

- Provide a detailed explanation of the ESG objectives of the fund and how the investment strategy encapsulates these objectives. Like article 8, the description must include which methodologies are established to assess, monitor and measure them.
- If the fund uses a benchmark index, a description from an ESG perspective is necessary. This is the discrepancy between the fund's investment strategy and the index composition. In addition, the funds must motivate why the fund applies a particular index versus more broad indices.
- If the fund does not use an index. In that case, it must provide a detailed explanation of its ESG objectives, and it may be necessary to describe why it differs from currently available indices in the market. We recommend that the investor is up to date on the launch of new indices, which may better fit the ESG objectives of the company.
- If the fund's objective is to invest in companies too, e.g. reduce carbon emissions, then an explanation is needed concerning the 2015 Paris Agreement.

In addition to the SFDR, there is currently a wide range of ESG labels that a fund can achieve to be used for marketing purposes (e.g. ecolabels).

Impact of SFDR

As established, we believe it is essential for the IRO to have an updated "wish list" of equity analysts and shareholders and embrace their scope in the IR tools. As the impact of MiFID II on equity research coverage, we think it is necessary to highlight the preliminary impact of SFDR. In 2021, the influential financial services rating company, MorningStar, has analysed the impact of SFDR from March 2021 to July 2021 in detail.[2] The key conclusions are:

- Funds and PMs have significantly updated their investment strategies to embrace the field of ESG and meet the requirements of articles 8 and 9.
- Actively managed funds will use articles 8 and 9 as of marketing and embrace the regulation much more than passive funds.
- Funds have established similar ESG investment strategies for articles 8 and 9 but differ in wide-ranging approaches.
- As intended by the regulation, articles 8 and 9 funds perform better on ESG metricise and PAI factors.
- Under MiFID II, investment firms must consider and integrate their clients' ESG preferences when engaging in investment product marketing and advisory.

Relevance for Non-EU Investment Companies

If an investment company markets towards end-investors in the EU, then the company is in the scope of SFDR. This is defined as AIFM directive[3] art. 42. These funds are required to disclose fund-level disclosures of PAI development and corresponding investment strategic updates on ESG.

The regulation has an indirect impact on "third-country" investment companies, especially for those non-EU PMs that have delegated investment mandates from an EU/EEA entity (common for Luxembourg/Liechtenstein setups). The non-EU PM must assist the representatives of investment mandates with SFDR compliance; this includes integration of ESG risks, consistency of remuneration policies and annual PAI reporting obligations.

The Interaction Between Taxonomy Legislation and ESG Reporting

In the next section, we will dive into how the IRO can embrace taxonomy. For both SFDR and the taxonomy legislation, PMs, PE managers and other asset managers have been required to disclose their investment strategies about ESG. It is encouraged that the ESG methodologies of articles 8 and 9 in SFDR are aligned with the principles of taxonomy. The entity must consider the proportion of ESG-related investments in its funds and how they're aligned with taxonomy and their PAI data. In the ongoing annual PAI reports, a detailed description of the progress must be considered regarding the fund's objectives. In respect of the MiFID II amendments, it will be necessary for the investment fund or entity to incorporate the relevant ESG methodology, the SFDR classifications, and the principles of taxonomy alongside their PAI data in order to consider their clients' sustainability preferences. Today, this remains a difficult task as there is a shortage of ESG data supplied by the companies in terms of both volume and quality. If the investment funds cannot obtain sufficient data and thus cannot align their investments with their clients' sustainability preferences, then they will gradually be forced to divest off their holdings in companies that are unable to disclose the necessary data.

A summary of facts and best practices is illustrated on page 329.

References and Sources

1. Schmucki, P. (2021). *The sustainable finance disclosure regulation is here*. KPMG.
2. Bioy, H., et al. (2021). *SFDR: Four months after its introduction, article 8 and 9 funds in review*. Morningstar Manager Research.
3. European Union. (2019). On alternative investment fund managers and amended directives 2003/41/EC and 2009/65/EC and regulations (EC) No. 1060/2009 and (EU) No. 1095/2010). *Official Journal of the European Union*. 2011/61/EU.

PART VII

The Framework of Best-in-Class Non-financial Reporting

CHAPTER 34

Implementing Taxonomy

It is not an investment if it is destroying the planet.

—Vandana Shiva, Indian scholar and Environmental activist

Framing the Taxonomy Legislation

EU countries and governments already make estimates of carbon footprint and other ESG criteria to determine a country's green economy. Different data sources are used, like administrative data, telephone calls and regular production surveys, and many companies provide data at a very detailed product level, i.e. bottom-up (Chapter 4). The taxonomy framework should assist in overall transparency by having the company's report the necessary data themselves and determine what sustainable economic activities is. In addition, it allows transparency for companies to report net-zero emission ambitions on different layers of their supply chain (so-called scope 1, 2 and 3).

As outlined in Chapter 16, the taxonomy legislation complements the CSRD (for non-financial regular companies) and the SFDR (for investment companies) by providing a common reference point for reporting the degree of alignment with sustainable activities. Despite the burdensome first-hand impression of integrating IR and ESG, we believe market participants should seize the opportunities offered by taxonomy to be at the forefront of non-financial reporting and strategically apply the framework to better their business models (Fig. 34.1).

P. Lykkesfeldt and L. L. Kjaergaard, *Investor Relations and ESG Reporting in a Regulatory Perspective*, https://doi.org/10.1007/978-3-031-05800-4_34

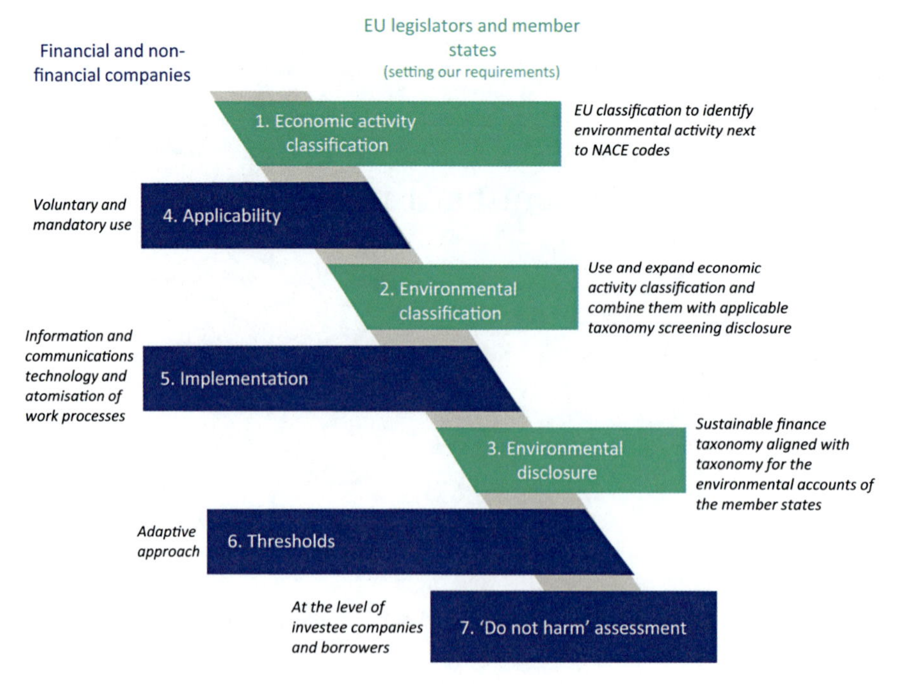

Fig. 34.1 Defining responsibility: companies and legislators (*Source* KPMG 2021)

Implementing Taxonomy in Non-financial Reporting

The first step is the most difficult in any transformative strategic process. It is demanding that companies collect data to evaluate their activities against the taxonomy framework and criteria. Next, companies must rethink strategies and information systems to feed the metrics and produce compliance information. This is especially true for required data points that demand actual measurement, not estimates. We believe, taxonomy is best implemented in three stages:

1. **Economic activity**: Identify each potential economic activity that the company's business model is exposed to. This includes the level 2 subcategories of transitional and enables economic activities.
2. **Eligibility**: Assess eligibility by comparing the company's revenue, OPEX and CAPEX to the technical standard of "substantial contribution".
3. **Reporting KPIs**: Report eligibility as the proportion of revenue, OPEX and CAPEX. Include contextual and transparent information on how the eligibility is determined and implemented in the necessary calculations.

Economic Activity

The EU has published the technical standards of taxonomy in a 600-page report.[1] The measures so far include the screening criteria for art. 10 ('Climate mitigation'), art. 11 ('Climate adaption') and art. 17 ('Not harm the other social objectives (DNSH)'), see Fig. 16.6 for the remaining screening criteria yet to be defined by the EU. The level of detail in the framework is astronomical. The company and the IRO must accurately filter the relevant information relative to its operations.

The EU has identified 8 of the 21 economic level 1 sectors that substantially contribute to climate change mitigation based primarily on carbon emissions. They have added two level 1 sectors for climate change adaption:

1. Forestry.
2. Agriculture.
3. Manufacturing.
4. Electricity, gas, steam and air conditioning supply.
5. Water supply, sewerage, waste management and remediation.
6. Transportation and storage.
7. Information and communication.
8. Construction and real estate.
9. Professional, scientific and technical activities (not included in "climate change mitigation").
10. Financial and insurance activities (not included in "climate change mitigation").

The company must identify which (if any) of the 68 economic level 2 subcategories of activities that its business model is exposed to. Please note that the framework is holistic and live screening criteria. Therefore, it will be adjusted and regularly refined basis.

Eligibility

Before commencing on non-financial reporting, we believe it is wise to familiarise itself with the economic categories most relevant for its business activities. Within each economic category, the EU has defined technical criteria of (a) core principles of taxonomy eligibility, (b) metrics and thresholds for eligibility and (c) do no significant harm assessment (DNSH) on other environmental objectives.

The technical criteria, tools and labels help identify sustainable activities, products, services or processes. Based on the company's economic activity, a company can line-by-line determine what sustainable or taxonomy eligible is and what is not. When companies disclose the percentage of sustainable activities, they might refer to the EU taxonomy and other sustainability standards for products or processes (Fig. 34.2).

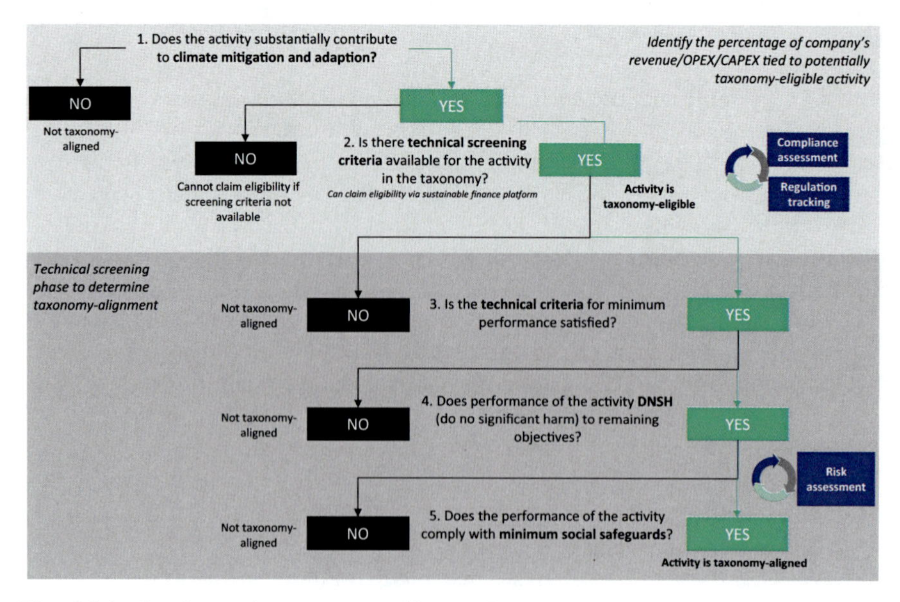

Fig. 34.2 Implementing taxonomy (*Source* European Commission, Goldman Sachs Global Investment Research)

Reporting KPIs

Companies are required to integrate the taxonomy factors into three main KPIs: revenue, CAPEX and OPEX:

- The revenue KPI represents the proportion of the revenue derived from products or services that are taxonomy eligible. The revenue KPI gives a static view of the company's contribution to environmental goals.
- The CAPEX KPI represents the proportion of CAPEX of an activity that is either already taxonomy-aligned or is part of a credible plan to extend or reach taxonomy alignment. CAPEX provides a dynamic and forward-looking view of companies' plans to transform their business activities.
- The OPEX KPI represents the proportion of the operating expenditure associated with taxonomy-aligned activities or the CAPEX plan. OPEX covers direct non-capitalised costs relating to research and development, renovation measures, short-term lease, maintenance and other direct expenditures relating to the day-to-day servicing of property, plant and equipment assets necessary to ensure continued and effective use of such assets (Fig. 34.3).

Many of the respondents that commented on the commission's technical standards have mentioned the challenge of collecting taxonomy compliance data

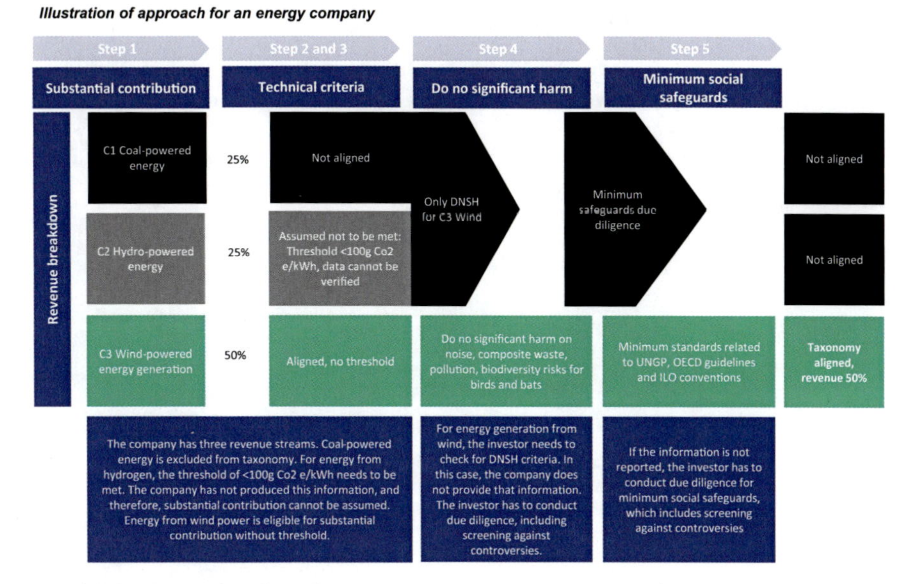

Fig. 34.3 Example of implementing taxonomy on revenue for an energy company (*Source* Baddache 2021)[2]

within companies and assigning this data to specific business lines. Therefore, there are critical attributes that the company needs to consider when calculating the taxonomy eligibility:

- The best judgement principle must be applied to separate their KPIs, avoid unduly inflation of taxonomy eligibility and avoid double-counting.
- The company must be transparent in the methodology in how the KPIs are prepared, what is included and assess the alignment to taxonomy.
- Disclosure should be provided by economic activity along with a total per KPI across economic activities at the level of the subsidiaries and group.
- The company should identify economic activities that are enabling (Article 16) and transitional (Article 10(2)).
- From IRO perspective, the communication towards investors and the taxonomy reporting needs to be aligned, especially in CAPEX plans.
- Ensure annual reporting and update data in line with the regulation's tracking, compliance and risk assessment.
- The company should integrate the taxonomy reporting process with general regulatory training and workshops.

References and Sources

1. EU, Technical Export Group (TEG). (2020). Taxonomy report: Technical annex: updated methodology and updated technical screening criteria. *Official Journal of the European Union*.
2. Baddache, F. (2021). *The EU taxonomy: A game changer for business and investors*. Ksapa. https://ksapa.org/the-eu-green-taxonomy-a-game-changer-for-business-and-investors/.

CHAPTER 35

Implementing Non-financial Reporting Standards

Scoping a Fragmented Market of Reporting Standards

It is a challenge combining the technical standards of taxonomy with a non-financial reporting framework. This is especially evident, given the missing harmonised principles of non-financial reporting—along with a fragmented market for ESG rating services. It therefore remains a complex topic for the financial community and the IRO. Finally, however, the company decides to implement a methodology. It should be clearly outlined, defined and explained in the disclosure.

Overview of Non-financial Reporting Frameworks

Currently, there is no single objective standard for measuring non-financial performance. This is presently being focused on, as the industry needs more standard principles of non-financial reporting standards.

- Global reporting initiative (GRI).[1]
- FSB's task force on climate-related financial disclosure (TCFD)[2] and appendix of implementation.[3]
- The climate disclosure standards board (CDSB).[4]
- International Business Council (IBC).[5]
- Standards of the sustainability accounting standards board (SASB).[6]
- International integrated reporting council (IIRC).[7]
- International sustainability standards board (ISSB), pending.

P. Lykkesfeldt and L. L. Kjaergaard, *Investor Relations and ESG Reporting in a Regulatory Perspective*, https://doi.org/10.1007/978-3-031-05800-4_35

Global Reporting Initiative (GRI)

Launched in Boston in 1997 (by non-profit organisations CERES and UNEP), the GRI is today headquartered in Amsterdam, and its standards are the most comprehensive non-financial reporting initiative globally. Widely used worldwide, references in national law are automatically included in some stock exchanges (e.g. automatically required on the Taiwan Stock Exchange and automatically in compliance with the stock exchanges guidelines on Oslo Børs, a stock exchange in Norway). GRI also administered a central sustainability disclosure database containing around 63,000 reports compiled over 20 years, but unfortunately, it was decommissioned in 2021.

Task Force for Climate-Related Financial Disclosure (TCFD)

The TCFD is a global initiative currently chaired by Michael Bloomberg. TCFD was created in 2015 by the Basel-based Financial Stability Board (FSB), which was set up in 2009 to promote international financial stability following the financial crisis. TCFD is not a regulatory body but has outlined a set of voluntary recommendations for non-financial reporting. They aim to make climate-related reporting disclosure more consistent and therefore more comparable. The objective is that better information allows companies to include environment-related risks and opportunities into their risk management, strategic planning and decision-making process. TCFD updated its framework of implementations in 2021.

Standards of the Sustainability Accounting Standards Board (SASB)

An independent non-profit born from responsible research at Harvard University, the organisation sees non-financial reporting as material sensitive (Chapters 2 and 3). It should therefore be included in the existing regulatory burden of financial reporting. The organisation sets a small number (77) of sustainable standards for industry- and sector-specific non-financial reporting.

International Integrated Reporting Council (IIRC)

Initially published in 2010 and updated in 2021, IIRC created an integrated framing framework (we choose not to use their abbreviation "IR") to integrate non-financial and financial reporting into a single disclosure performance framework. It does not distinguish between ESG factors, as IIRC sees them inherently intertwined. Instead, the framework considers a) business model considerations, b) responsibility for an integrated report and c) charting a path forward.

The World Economic Forum (WEF)

The World Economic Forum (WEF) founded the climate disclosure standards board (CDSB) in 2008. The CDSB framework focuses entirely on climate reporting, energy strategy and climate change.

Considering the focus on consolidated non-financial reporting, the WEF set up the International Business Council (IBC) in January 2020. The project was to define standard metrics for sustainable value creation, the aim being to improve the ways that companies measure and report on their contributions towards more prosperous, fulfilled societies and a more sustainable relationship with the planet. In September 2020, the IBC published its core Stakeholder Capitalism Metrics and disclosures. Companies can use these to align their mainstream reporting on performance against ESG indicators and consistently track their contributions towards the SDGs. The metrics are organised under four pillars aligned with the SDGs and principal ESG domains: principles of governance, planet, people and prosperity.

Recently, at the COP26 UN global summit in November 2021, the IBC, WEF and the international financial reporting standards foundation (IFRS) announced the formation of the International Sustainability Standards Board (ISSB). ISSB will develop a global compressive baseline of high-quality sustainability disclosure standards.

Under ISSB, a technical readiness working group (TRWG) lays the technical groundwork for a global sustainability disclosure standard for the financial markets and gives the ISSB a running start. The TRWG will include representatives from the organisations mentioned above, i.e. WEF, CDSB, IBC, IFRS and ISSB. In addition, TCFD will also join the working group and other organisations.

Non-financial Reporting Framework: Why Is There More Than One?

The framework is a fragment, but the standards have high similarities. For example, CDSB is specific for environmental reporting, whereas GRI, TCFD, IBC, SASB and IIRC integrate social and corporate governance factors. The latter five have all publicly announced that their frameworks complement each other and are designed for different but complementary purposes. In addition, they have alignment between ESG categories and methodology:

- **GRI's** standards support sustainability reporting and are based on the company's ESG impacts and contribution (positive and negative) towards sustainable development. The standards focus on helping organisations communicate their impact outwards related to ESG factors and how they are managed and impacted.
- **TCFD** is based on transparent financial risk disclosure recommendations and a tangible framework for setting non-financial reporting goals. The

TCFD focuses specifically on helping organisations disclose the financial impacts of climate change risks and opportunities.
- Where **GRI** standards support comprehensive disclosures and look at the company's impacts on the world, **SASB** attempt to identify the world's impact on the company.
- **IIRC** is an alternative reporting method, not distinguishing between financial and non-financial data or ESG factors.
- **IBC** is WEF initiated and sponsored by the big-four auditing firms Deloitte, EY, KPMG and PwC. The standards are well-aligned with the SDGs. They complement the GRI, TCFD and SASB considerations.

Practically Navigating the Fragmented Market

WEF's ISSB will include TCFD and IBC in laying the technical groundwork for a global sustainability disclosure standard setting. These will be implemented by the ISSB, which will provide one unified framework that encompasses all the above standards.

We are not aware of a projected date of publication. However, we expect the framework merely to be highly aligned with the GRI, TCFD, SASB and the SDGs. We would like to highlight the valuable IAS Plus online resource[8] published by Deloitte; that we found helpful to track the ongoing changes and amendments within the field of non-financial reporting.

The field is emerging, but in terms of practical assistance, Nasdaq published in 2019 a helpful guideline to integrate the necessary standards and frameworks listed above. We have attached an overview in the appendix, for your convenience.

The ESG reporting guide is intended to serve as a resource to increase operational efficiency, decrease resource dependency and attract a new generation of empowered workers. We have listed the main takeaways from the different ESG categories in the following three chapters, which we recommend any IRO to become familiar with. This overview is intended to help understand which resources are necessary to mobilise to produce non-financial reporting. The categories are a combination of quantitative and qualitative measures, and a company should ensure to apply the "comply and explain" principle to all of the categories.

A summary of facts and best practices is illustrated on page 329.

Appendix

See Fig. 35.1.

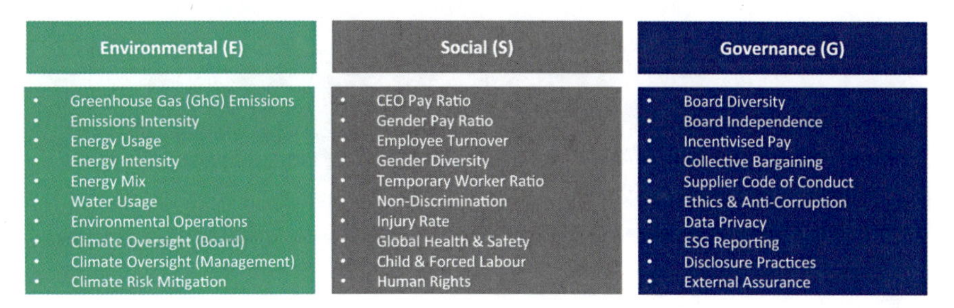

Fig. 35.1 Overview of ESG sub-categories of disclosure[9] (*Source* Nasdaq [2019])

References and Sources

1. Gri. (2021). Full set of standards of GRI standards. www.globalreporting.org/how-to-use-the-gri-standards
2. Task force on climate-related financial disclosures (TCFD). (2017). Recommendation of the take force on climate related financial disclosures. *Final report.*
3. Task force on climate-related financial disclosures (TCFD). (2021). Implementing the recommendations of the 'TCFD'. *Final report, superseding the 2017 annex.*
4. Climate disclosure standards board. (2019). CDSB framework: for reporting environmental and climate change information, advancing, and aligning disclosure of environmental information in mainstream reports. *CDSB:* https://live-cdsb.pantheonsite.io/sites/default/files/cdsb_framework_2019_v2.2.pdf
5. WEF. (2020). Measuring stakeholder capitalism: towards common metrics and consistent reporting of sustainable value creation. *World economic forum.*
6. SASB standards. (2018). SASB's 77 industry standards and the sasb standards application guidance. www.sasb.org/standards/download/
7. IIRC. (2021). International integrated reporting framework. https://www.valuereportingfoundation.org/wp-content/uploads/2021/07/InternationalIntegratedReportingFramework.pdf
8. *Link:* https://www.iasplus.com/en/resources/sustainability
9. Nasdaq. (2019). ESG reporting guide 2.0: a support resource for companies. *Nasdaq.*

PART VIII

Preparing the Company's First ESG Report

CHAPTER 36

For the First Non-financial Report: A Checklist to Get into Gear

> The SDGs are describing the world we all want. A world which is very different from the one we know today. And in order to get there, it requires the biggest transformation mankind has experienced so far (…) This is why we need visionary and holistic business leaders who can translate the Global Goals into business strategy.
>
> —Lise Kingo, Independent Board Director in Sanofi SA, Covestro AG and Aker Horizons ASA, and former CEO and Executive Director of United Nations Global Compact.

Setting the Agenda for the company's First Non-financial Report

Strategically embracing non-financial reporting and stakeholder capitalism (Part VI) comes from the top-down, recognised by the UNGC. Therefore, senior management and the IRO must have the board of directors to formally take the initiative and to set the ESG agenda.

It is very important that the selected ESGs are anchored at the strategic level in the company's purpose, governance and business strategy. The board of directors must then subsequently be updated regularly on the implementation, progress and feedback. It is recommended to include the board of directors as the initial stage of framing the approach.

Before commencing on designing metrics and targets based on taxonomy and CSRD of relevant principles, the company needs to identify the most appropriate framework. The company is recommended to commence by identifying the most material SDGs (typically 4–5) in respect of its business and

P. Lykkesfeldt and L. L. Kjaergaard, *Investor Relations and ESG Reporting in a Regulatory Perspective*, https://doi.org/10.1007/978-3-031-05800-4_36

then frame the ESG report around the selected SDGs as well as consider the science-based targets (SBTi) in relation to climate change. The SBTi website has a valuable guideline on choosing and preparing the company's approach to SDGs. The company should also consider their peers and include relevant stakeholders, senior management and the board of directors.

After selecting the most relevant SDGs, the company must frame and work with its primary audience. In our view, the most critical consideration for the IRO is to work with internal stakeholders to make the job of the investment community and stakeholders as straightforward as possible. That means high transparency, coherency, availability and comparability.

Aspire to Integrate Best Practices from the Beginning

We have noted that the annual report is the most critical tool for any IR (Chapter 21). It should be the foundation of investor communication and the most essential source for the financial community to receive complete and comprehensive information. Likewise, it is also the most critical source of information for the company's primary stakeholders—including its employees. They must be considered when setting and tracking the company's non-financial goals. The company needs to ensure consistency by following international reporting standards.

Today, most companies publish a separate ESG/sustainability/responsibility /CSR report to complement their annual report. However, we essentially suggest integrating the two reports (which we believe with be the trend over time), or at publishing the reports simultaneously. A company must ensure regulatory compliance and allow the financial community and the stakeholders to be aware of the company's ESG strategy, initiatives and progress. Therefore, it is critical to publish key ESG figures with key financial figures in the annual report and perhaps include granularity in the ESG report.

The annual report and the ESG report must not have significant differences. In many EU member states, the financial authorities require every company to include non-financial data (e.g. CO_2 emissions, diversity and sick absence) in the annual report.

Most non-financial reporting figures complement the financial figures (Chapter 7) and highlight a best-in-class ESG approach. Notably, the P&L is positively impacted by the efficiency of energy costs and employment churn. However, energy costs and employment churn have varying importance depending on its business model and sector. Therefore, the relevancy of ESG figures vary. Therefore, a company should communicate how the various ESG criteria impact stakeholders and to what extent (Fig. 36.1).

Class	Category	Sub-category (examples)	Employees	Community	Suppliers	Customers	Shareholders
Environment	Carbon and climate	Energy and fuel efficiency GhG emissions Technology and opportunity (investments)		✓	✓		✓
	Natural resources	Water (use and pollution) Land, forests, biodiversity (use and pollution) Sustainable sourcing		✓	✓		✓
	Waste and toxicity	Hazardous and non-hazardous waste Emissions and spills Electronic waste Packaging material	✓	✓	✓	✓	✓
	Management of environmental risk	Disaster planning, response and resiliency LEED design and certification	✓	✓	✓	✓	✓
Social	Human rights	Ethical sourcing Supply chain standards	✓	✓	✓	✓	✓
	Labour, health and safety	Fair wages, benefits, training and development Labour standards, job stability and mobility Employee engagement	✓	✓	✓		✓
	Diversity and inclusion	Equal opportunity and participation	✓	✓	✓		✓
	Product safety, quality and brand	Customer satisfaction Affordability and accessibility	✓	✓		✓	✓
	Community engagement / partnership	Volunteer hours Workforce/community demographic parity Alliances with key stakeholders Corporate philanthropy	✓	✓			✓
Governance	Board composition	Minority representation Gender equality	✓	✓		✓	✓
	Ethics and compliance	Anti-corruption Cybersecurity and data privacy Oversight and accountability Management policies, systems and disclosure (transparency) Political contributions/lobbying	✓	✓	✓	✓	✓
	General corporate governance	Senior management compensation Board leadership/structure Share structure (multiple classes, board election)	✓				✓
	Risk management and mitigation	Code of conduct Sensitivity analysis and stress testing	✓	✓		✓	✓

Fig. 36.1 ESG criteria and stakeholders

Integrate the ESG Process with Financial Reporting

Like financial data, non-financial data must be highly reliable. Adding credibility by having the numbers independently audited is recommended consideration and can then be put into context with financial data. We recommend always putting ESG figures into context. How are the figures defined, developed and the targets? All of this results in credibility in the financial markets.

In addition, providing "building blocks" to achieve the targets is a valuable feature for financial reporting and can also be utilised. For example, including considerations behind the incentive and remuneration structure (Chapter 37) may assist stakeholders and the financial community understanding how the company plans to implement the targets and to develop onwards.

Sometimes providing data that is accurate, relevant and harmonised is impossible. For example, if a company chooses to focus on the first two factors and provide normalised data relating to the company financial reporting. In that case, the reader should find the raw data in the footnote, or in the appendix, to ensure harmonisation. This allows comparison between the company and its peers. The key foundation for the first non-financial report is:

- **Reporting boundaries:** Data map all variables (from SGDs to other non-financial reporting metrics) and always use the same reporting boundaries and standards as for financial reporting.
- **Consolidation:** Always follow the financial reporting standards for consolidating data and apply standardised measurements.
- **Period:** Always follow the periods of financial statements.
- **Performance and trends:** Assure forward-looking quality and relevant information on challenges and opportunities. However, it also clearly explains the historical performance for the past 3–5 years. Include use cases on a macro- and micro-level on how the performance will develop to meet the company's targets.
- **Accounting standards:** Always disclose the accounting standards used in the non-financial report and onboard an external assurance company in due time.

Strategically Framing the company's First Non-financial Report

We are finding better ways to measure and report performance as ESG data is growing even more relevant and vital. However, the field is not yet mature and therefore continues to develop. Consequently, it is imperative to frame the company's first non-financial report but still continuously improve it strategically. Given the impact of the CSRD (Chapter 32), it is increasingly important for the company to mobilise resources to put non-financial reporting on the agenda. CSRD reporting ensures the company maintains access to capital, ensures profitability and growth, complies with legislation and risk management, upkeeps corporate reputation towards stakeholders and improves transparency (and at the same time lower the risk premium associated with the company's shares).

Following the implementation of the UNGC's SDGs implementation framework (Part VI) and considerations on taxonomy (Part VII), we believe a company can strategically frame a non-financial reporting approach through the four Es of **establish, expand, embed and enhance.**

- **Establish a foundation: The company must build a tangible framework for non-financial data :**
  - Identifying and defining risks, opportunities and cost–benefit on the degree of non-financial reporting from a strategic and investment perspective (Part I).
  - Considering the role of other non-financial markets stakeholders (Parts II and III).

- Integrate non-financial reporting into internal resources, including the IR department (Part IV) and apply IR tools efficiently, e.g.:
    - Integrate with regular IR tools.
    - Include the company's ESG representatives in regular meetings, so that all senior managers are able to communicate their expertise feedback.
    - Benchmarking of non-financial reporting themes relative to the company's peers and competitors.
    - Include the company's stakeholders and non-financial reporting themes in IR perception studies.
    - Integrate non-financial reporting with digital IR.
- Maintain a robust corporate governance and IR set-up and consider non-financial reporting on par with financial reporting in special situations (Part V).
- Create internal workshops and sessions on the importance of mobilising and framing non-financial reporting in business strategies, especially with the focus on being a listed company (Part VI).
- Design a preliminary roadmap for the non-financial reporting journey: Before commencing step-by-step guidelines.

- **Expand : Based on an established framework, a narrowing approach to key principles is identified (Part VI):**

    - Study the official reporting frameworks of ESG guidelines and scope their importance relevant to the company's strategic direction and non-financial reporting aspirations.
    - Identify most relevant non-financial reporting principles to the company and its strategic objectives, product/services and operations.
    - Scenario analysis and scoping of the identified principles.
    - Upskilling and training for senior management and board.
    - Design metrics and targets of the identified relevant principles.
    - Commence on internal reporting and communications procedures.
    - Conduct analysis of taxonomy (Chapter 34).
    - Transform and update the existing incentive structures to encompass and mobilise non-financial reporting (Chapter 37).

- **Embedding of the reporting: A long-term design and implementation are required based on an established framework:**

    - Implementation of an internal roadmap for non-financial reporting.
    - Involve, communicate and implement incentive structures to line managers.
    - Implement the decarbonisation and strategy of new principles.

  - Monitor non-financial reporting risks.
  - Create governance support for relevant stakeholders.

- **Enhance the quality: For the embedded non-financial reporting structure, a company should achieve a best-in-class approach (Part IX)**

  - Include stakeholders and non-financial reporting themes in perception studies.
  - Ongoing benchmarking to peers and best-in-class non-financial reporting firms.
  - Engage and maintain collaborations with stakeholders.
  - Collaborate and monitor rating agencies.
  - Best-practice IR communication: Include non-financial reporting into IR strategies and including equity analyst and shareholder targeting.

- Revert to the company's basic principles and mobilising stage in the "establish stage" (Fig. 36.2).

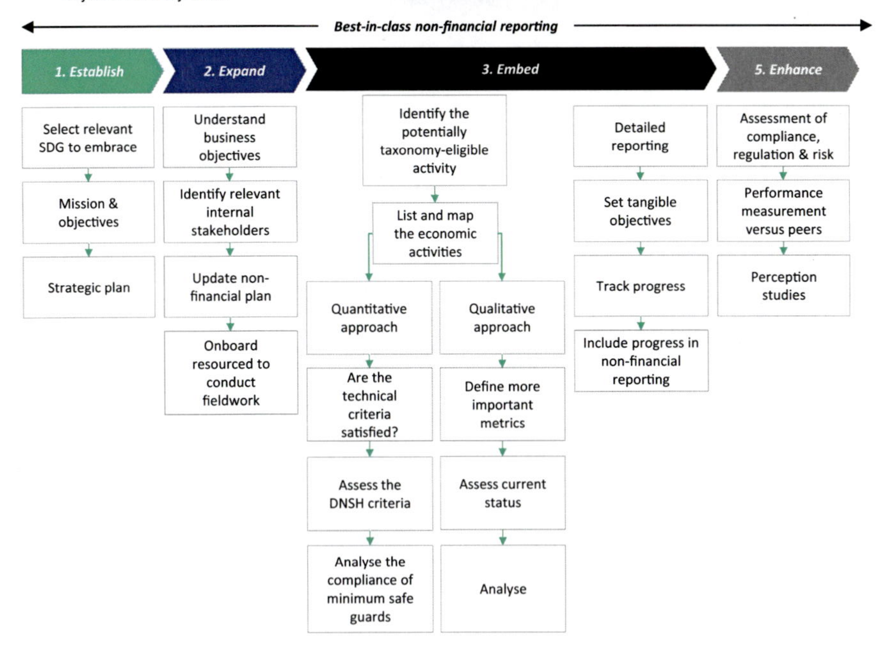

Fig. 36.2 Best-practice non-financial reporting process

CHAPTER 37

The Importance of Mobilising ESG with Incentives

Mobilising the Company to Set ESG Goals

By now, it should be evident that ESG must be integrated into the business strategy and not be an independent focus area. Therefore, the employees that serve as the company's most important stakeholder must be involved in a successful implementation. With an integrated strategy, the company is also recommended to have a balanced business scorecard that relates to both the financial and non-financial priorities. Today, we suspect most variable remuneration is based on reaching financial quantitative targets.

There is a common phrase within incentive structures: "You get what you measure", i.e. if a company only decides to track and incentivise on revenue performance, then sales of the company will likely go up. However, there is a risk that the company will have unfavourable onboard contracts, as this is not incentivised. On the other hand, if a company decides to track and incentivise free cash flow, then sales will likely not increase as much. Still, the onboarded contracts will probably be deemed sustainable from a cost–benefit analysis.

People respond to incentives. Without an in-depth analysis of the massively important topic of motivating people, besides receiving a paycheque, increasingly employees (especially considering COVID-19) see "purpose" as a primary motivating factor for meaningful work.[1] In our experience, most business strategies fail if a company or leader fails to mobilise the stakeholders and involve them in the process.

If stakeholders are not committed to the purpose of a strategy, then the motivation for change is limited. Therefore, it is essential to align internal stakeholders (board of directors, senior management and senior general staff) and mobilise them to set ESG goals and associated incentive goals. Mobilising purpose and implementing incentive goals will ultimately drive change.

P. Lykkesfeldt and L. L. Kjaergaard, *Investor Relations and ESG Reporting in a Regulatory Perspective*, https://doi.org/10.1007/978-3-031-05800-4_37

We believe that including quantitative and qualitative targets related to both financial and non-financial priorities sends a very strong signal to the organisation that everyone is contributing to the same goals. Even though some have more tangible impact, for example a factory manager versus the factory employees, it is important to involve everyone and include them in the overall purpose—but not alone be responsible for the goals.

Aligning Incentives with Non-financial Reporting

ESG incentives are like traditional incentive measures; they should support and reinforce strategy and direction. As a result, the IRO should embrace relevant ESG goals and objectives and consider a company's quantification of specific ESG issues. Besides initiating the process, companies should evaluate their readiness of adopting these goals through the following framework:

- **Companies developing an ESG strategy:** This is relevant for the preliminary stages of developing an overall non-financial/ESG reporting strategy. These companies should identify, frame, set and initiate their ESG framework before implementing incentive structures.
- **Companies ready to set qualitative ESG goals:** This is relevant for companies that have completed the initial stages of initiating their ESG framework and can now formalise the reporting disclosure to their stakeholders. These companies should have implemented primary goals on ESG factors, including diversity, renewable energy, emission goals, etc.
- **Companies ready to set quantitative ESG goals:** This is relevant for ESG-developed companies with a clear purpose and established reporting framework on qualitative ESG goals to now outline, report and strategies on quantitative goals, including carbon reduction targets, net-zero commitments, diversity and inclusion, along with matrices such as customer and supplier satisfaction. As we will explore, these companies should strive for best-practice IR and best-in-class ESG reporting.

No matter which readiness stage a company is in, it is vital always to strive to increase value to its identified stakeholders. Simply implementing an ESG strategy with no purpose will be seen as "greenwashing" and "window dressing". This is especially true for including incentive structures on ESG factors that have no inherent benefit for a company.

To illustrate the point, it is much less relevant for a software company to base its ESG incentive structure on plastic waste than reducing its electricity consumption. Likewise, it is less appropriate for a financial service company to focus on lowering its waste consumption than on loans to minority groups, inclusion statistics or perhaps anti-money laundry (AML) targets.

Aligning ESG with the Company's Value Creation

"Greenwashing" and "window dressing" are unfortunately not uncommon trends for non-financial reporting in the financial markets. We believe communicating relevant and transparent information with associated objectives and a clear incentive structure increases trust in the company's ESG and value creation drastically. Creating goals is especially tough, as they much are aligned with the company's overall value creation but also consider the company's stakeholders:

- **Employees:** Like an IR perception study, a company is also recommended to survey its employees on which non-financial reporting considerations they value highest, social initiatives? Environmental considerations? The company can use the survey to frame their overall sustainability targets and communicate their progress to employees. This is a helpful tool to enhance employee engagement.
- **Customers:** What ESG considerations do the customers have? The industry, sustainable supply chains, social justice and environmental footprint can weigh differently.
- **Long-term sustainability:** If considering its strategic function, how do they view the overall macro-factors? Rising legislation or prices on certain input materials are essential considerations shaping targets and framing CAPEX.
- **Brand Image :** What is the company missing a step back from the above considerations? The marketing term "think global, act local" can wisely be applied. For example, a company may decide to build a state-of-the-art carbon neutral factory with best-in-class water consumption, a high focus on local minority recruiting and a decentralised incentive structure. However, if it ruins the view of the local town's houses, all the above may seem redundant.

Like financial reporting and KPIs, there are many standards formulas, terms and calculations to be applied—however, it must be tailored to the company's short- and long-term care. For this reason, considering the company's stakeholders and their size, industry, motivation and relationships is essential before designing an ESG incentive structure.

Including ESG in the Design of the Incentive Structure

Like setting standard incentive structures, the objectives must be specific, measurable, achievable, realistic and time-related (also known as "SMART").

Designing an incentive structure needs to be associated with the company's short-term goals and long-term ESG objectives. In addition, the company, including the IRO and the external communications manager, needs to be aligned on several criteria that may be implemented by the company and communicated to the financial community:

- **Quantitative goals:**
    - Judging by the relevance, quality and accessibility of ESG-related data, implementing concrete incentive structures can prove more difficult than financial measures.
    - For now, we believe it is imperative to include quantitative ESG data to complement existing incentive structures.
    - It is also almost impossible to track ESG data on a personal employee level compared to financial measures. Therefore, it can be more relativistic to use the metricise on a group, e.g. an office, department, country or BU.
- **Qualitative milestones:**
    - In our experience, many incentive structures have multiple qualitative components which an employee's line manager broadly contemplates.
    - Including ESG-related information and specified milestones can assist in driving the company's overall objectives.
- **Selective relevant metricise:**
    - Various ESG metricises and data points mean differently to companies depending on their location, industry, operating model, stakeholders and purpose.
    - A company can utilise a principles-based approach and assess the "SMART" framework to find the more relevant metricise.
- **Employee participation :**
    - No company should ever give incentive targets to an employee who has no control over the outcome. This can seriously hurt employee morale. As a result, we argue it would be wise to only include line managers and senior managers in the scope of the incentive structure.
    - If the company decides to track and compensate groups on their quantitative ESG development, including all layers of the hierarchy, it could have positive ripple effects.
- **Applying "weights" to achieve goals:**
    - Compared to the US, the relative size of variable compensation schemes and incentive structures is limited. We estimate that relatively few (non-sales related) line managers receive more than 20% of their total compensation as a variable bonus.

- The variable has different categories with different weights that determine the level of variable incentive. Currently, we believe most of the individuals in the financial community, who have a bonus as a key element of their total compensation, would react negatively if more than 10–20% of the variable compensation would be directly ESG related.

- **Periodic (annual and long-term) goals:**
    - Some ESG targets may be better for long-term incentives than annual incentive compensation. For example, some milestones and performance measures are relevant annually; some need a longer horizon.
    - For example, reducing CO_2 consumption of a factory is useless on a one-year basis if the company plans to install carbon filters the following year.

- **Encapsulating ESG in long-term plans:**
    - Three-year performance incentives should be aligned with the group targets.
    - The long-term incentive plan could be based on free cash flow, revenue growth and ESG-rated targets.
    - The weight could be, e.g.:
        - 60% free cash flow (50% short term, and 50% long term, FCF).
        - 20% revenue growth (50% large contracts and 50% on small contracts).
        - 20% ESG-related targets (split into 50/25/25% on specified E, S and G factors).
    - The incentive could be a cash-bonus, restricted stock/share units (RSU) and tangible goods like traditional payment structures.

A summary of facts and best practices is illustrated on page 329.

Reference and Source

1. Badubi, R. (2017). Theories of motivation and their application in organizations: A risk analysis. *International Journal of Innovation and Economic Development*, Vol. 3, Issue 3, August 2017, pp. 44–51.

PART IX

Aiming for Best-in-Class ESG Reporting

CHAPTER 38

Taking Already Implemented Non-financial Reports a Step Further

But shouldn't all of us on earth give the best we have to others and offer whatever is in our power?

—Hans Christian Andersen, Author

Integrating Non-financial Reporting is an Art

A company should strive to deliver solid and transparent financial and non-financial performance while having a best-practice IR approach and a coherent equity story. If this is successful, then a company is expected to have a lower risk premium, lower its cost of capital and find it easier to raise new capital.

Satisfying existing (and potential) shareholders and stakeholders is expected to attract long-term institutional investors and superior equity analyst coverage. Moreover, if this mindset is embedded and practised in the long term, then the availability and magnitude of raising new capital will be easier.

Historically, the financial markets industry is a very social setting (Chapters 2 and 3). Therefore, storytelling is key in any analyst report, a pitch for an investor meeting, or communicating to the financial community. In our experience, the most respected listed companies are not only those that deliver solid financial performance but are also able to deliver transparent equity story and prober disclosure of financial figures.

Communicating an equity story towards investors and equity analysts is key to create exposure and interest. As established (Chapter 17), the foremost goal of the IRO is to ensure a transparent and coherent equity story, so the financial community understands the company's vision, mission, strategies, and business and financial conditions.

P. Lykkesfeldt and L. L. Kjaergaard, *Investor Relations and ESG Reporting in a Regulatory Perspective*, https://doi.org/10.1007/978-3-031-05800-4_38

The IRO should strive to establish a clear equity story between its strategic and operational aspirations and its financial reporting. As well as considering the role of other stakeholders, the IRO should strive for similar considerations in connection the non-financial reporting. This today is the most significant challenge of any IRO.

As non-financial reporting becomes embedded, the next step is to decide on the ambition level (Chapter 18) to continue to enhance the company's focus on its equity story and non-financial figures. This includes the interlink between its stakeholder approach and its strategic and operational performance. In addition, the IRO should strive to engage and maintain collaborations with stakeholders and employ the best-fit IR tools (Chapter 21).

Asking Questions is the First Way to Begin Change

They enhance a company's standing on the financial markets, including perception studies for different stakeholder groups. To ensure the best possible feedback, the IRO must maintain and engage with stakeholders—similarly to the approach to shareholders. Listening to investors, equity analysts and stakeholders will shape the direction for implementing and maintaining up-to-date non-financial reporting.

The IRO should be an expert in managing the expectations of the financial community (Chapter 22). Dialogue and feedback from the financial community allow the IRO and company to be at the forefront of the market perception of the company. Including all stakeholders in the regular dialogues will assist the company in proactive best-in-class ESG reporting.

Like marketing, we believe the amount of IR communication and non-financial reporting integration should be exceptionally high. However, customers can feel tired of constant marketing from a company, and likewise, the financial community can become tired of constant IR communication. Therefore, it is important to evaluate a company's standing concerning investors and equity analysts.

The brokers in investment banks have faced diminishing commission margins and research payments (Chapter 14). But, despite equity analysts and brokers losing power relatively to investors, they are by no means powerless. On the contrary, they continue to serve as the most influential external partners a company can have in the financial markets. Moreover, as the investment banks have an extensive list of investors, they can bring close attention to investors.

Equity analysts and brokers attempt to stand out in the market by distinguishing themselves. A great leap for an IRO is to develop a two-way relationship with selected equity analysts to understand how the market perceives the company at any point in time.

Drafting a Tailored Non-financial Reporting Policy

It should be clear that the approach to enhancing non-financial reporting is like the ongoing work of the IRO, which can deploy many of the same tools explored in this book. Once the company has established and expanded its approach to non-financial reporting, we believe the IRO and company management should draft an operational policy to ensure best-in-class non-financial reporting. The policy can be similar or integrated with the IR policies (Chapter 21.) and be tailored to the company business model, objectives and ambition level. For inspiration, we believe the topics could include:

- An updated list of key SDGs and how the equity story is aligned.
- Ongoing tracking of the ESG performance relative to peers and global objectives (including SDGs).
- Up to date on new trends in non-financial reporting principles and policies.
- Create ongoing perception studies among stakeholders (including investors, media, etc.). Include external ESG consultants to make the perception study relatively frequent.
- Include screening criteria on ESG investment strategies to improve investor and equity analyst targeting.
- Set and maintain qualitative and quantitative benchmarking on non-financial reporting objectives and the IRO's overreaching objectives.
- Set a reminder to listen and learn from ESG rating agencies, external stakeholders and NGOs.
- Ensure transparent areas of responsibility for the IRO, communication and other functions. Increase staffing in the ESG function if necessary.

Teachings from Service Giant, ISS A/S, on Taking the First Steps of the Full Corporate Integration of ESG and Non-financial Reporting

ISS A/S is the world's largest facility and service management company. Founded in 1901, its core services are cleaning, technical, catering, security and workplace management. The company employs around 400,000 people and is thus one of the largest employers in the world. Following a significant IT incident, the COVID-19 pandemic and several troubled contracts, ISS made significant changes to the senior management in 2020 and initiated a turnaround plan until 2022. The company seems to be on track with its turnaround. The company is now focusing on the next step of its journey.

Based on the TCFD recommendations, ISS in 2021 carried out a comprehensive review to identify and understand the climate-related risks and opportunities. The review included the company's governance, strategy, risk management and metrics and targets. It resulted in the development of a three-year roadmap for implementing the recommendations regarding climate-related risks and opportunities. We believe this 2022 update from ISS is a strong signal that other global leaders should also integrate its financial and non-financial considerations and align them across the organisation (Fig. 38.1).

As a result of ISS' review, the company has committed to several climate-related targets, which aim to develop sustainable business strategies, promote best practices in emissions reductions and mitigate the risks of climate change. As a result, on 26 January 2022, ISS announced that it aimed to reach full-scope net-zero greenhouse emissions by 2040. This includes net-zero scope 1–2 by 2030 and net-zero full-scope by 2040. To achieve this, the company plans to reduce its food footprint by 25% in 2030, halve the amount of food waste by 2027, electrify the global fleet of approximately 20,000 vehicles and commit to significant water reductions and usage of renewable energy. The targets will be verified and aligned to the SBTi.

To kick-start the ambitions, ISS has five concrete priorities for 2022. Two out of five (40%) are non-financial considerations concerning how ESG can create value for the company and its stakeholders (Fig. 38.2).

In the above-mentioned review, ISS included three significant considerations. Firstly, in line with UNGC's considerations, change management is driven top-down. Secondly, the company must invest in resources to drive

Recommendations from the Task Force on Climate-related Financial disclosure (TCFD)	ISS' initiatives
Governance Disclose the organisation's governance around climate-related risks and opportunities.	• At ISS, the board of directors to be ultimately responsible for risk, inclusive climate-related risks. • Executive management is responsible for sustainability and has established a Sustainability Committee addressing ESG-related matters, including climate-related risks.
Strategy Disclose the actual and potential impacts of climate-related risks and opportunities on organisation's businesses, strategy and financial planning where such information is material.	• The strategic ambition is to be recognised as an environmental leader, advocating for more sustainable actions, measures and goals. • ISS focuses on reducing the impact on the environment and climate contributes to solving the global challenge of climate change and creating a sustainable world for future generations.
Risk management Disclosure how the organisation identifies, assesses and manages climate-related risks.	• The three-year roadmap and long-term initiatives and activities are carried out systematically. The company identifies the potential for more efficient use of resources, lower emissions and cost optimization. • In addition, ISS aims to proactively mitigate environmental risk and anticipate the customers' needs.
Metrics and targets Disclose the metrics and targets to assess and manage relevant climate-related risks and opportunities where such information is material.	• Targets (%) for electricity and water consumption as well as car emissions. • Science-based targets (SBTi) initiatives (scope 1-3 emissions) net-zero targets. • Commitment to SDG-13 – ISS takes urgent actions to combat climate change and its impact to reach 2040 net zero. • UN Global Compact commitment. • Diversity and inclusion (D&I) targets. • Committed to TCFD reporting.

Fig. 38.1 ISS' response to the TCFD's recommendations

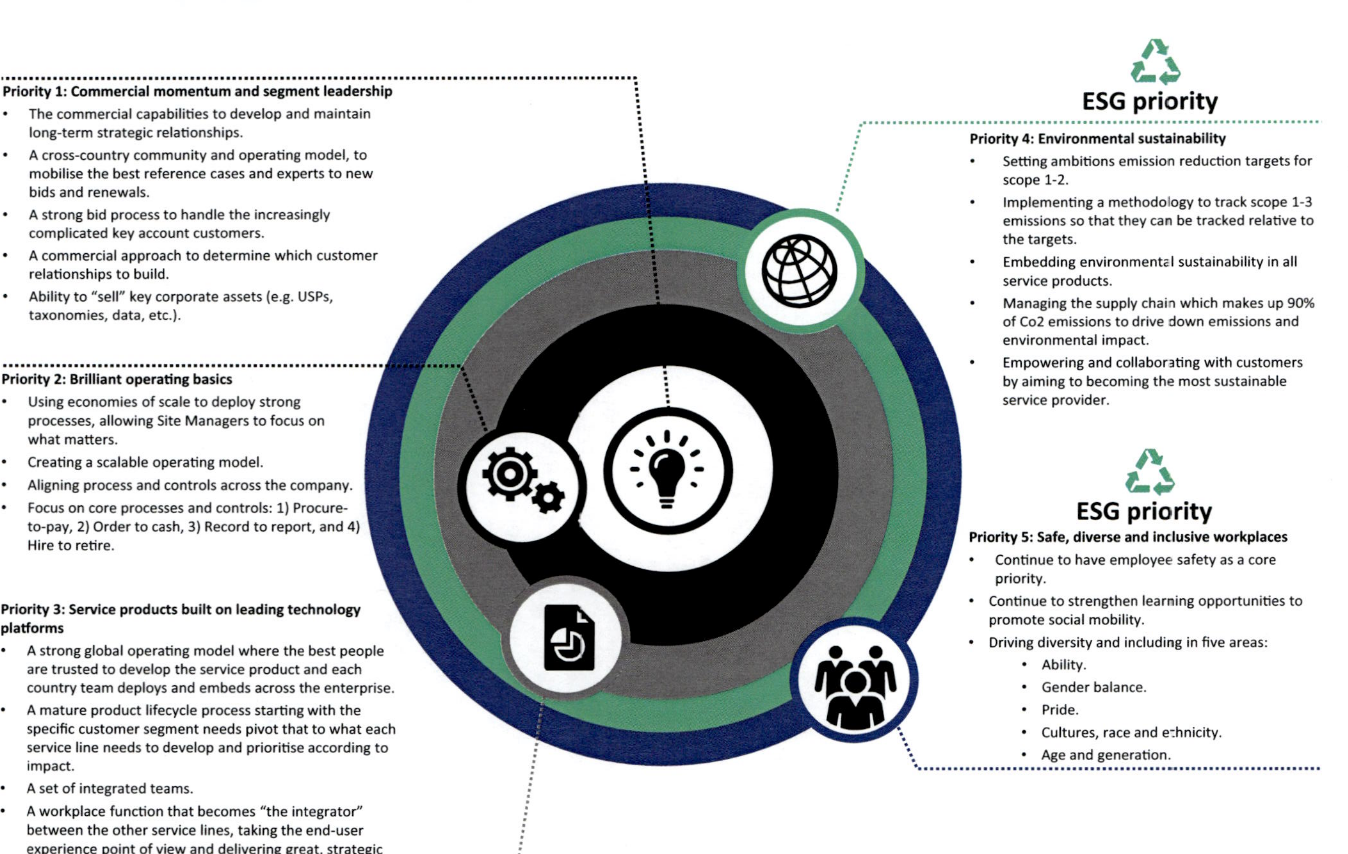

Fig. 38.2 ISS' five key priorities for 2022

the change, and thirdly, proper reporting and tracking measures must be implemented:

1. ESG to be integrated with strategy—led and guided from the top
   - Materiality assessment: Where can ISS make a difference to society, and what makes a difference for ISS. Relative more significant inherent focus on ISS' societal impact on society.
2. Investing resources and establishing an organisation
   - The transformation requires skills, so ISS has divided the opportunities into three functions:
   - Sustainability (people, planet, prosperity, governance).
   - Diversity and inclusion.
   - ESG reporting.
3. Reporting and tracking
   - ISS follows the Danish Chartered Financial Analyst Association's (CFA) recommendations for reporting and company-specific targets on ESG.
   - The company has initiated a separate organisation under the finance operation, conducting an internal audit on all disclosed figures.
   - ISS' auditors will conduct an audit of all non-financial/ESG figures.

We believe this is an example of a proactive and structured approach, initially inspired by the board of directors and then subsequently driven by executive management, regarding the first steps of non-financial considerations and implementation at the global level, which many companies can aspire from.

CHAPTER 39

Encompassing ESG Rating Agencies

ESG as a Rating Tool

All countries and even major companies have credit scores dictated by the institution's payment history, outstanding debt, credit history length, credit mix and the pursuit of new credit to discover the company's ability to make future payments. Unfortunately, there is no formal ESG score today and the market for rating agencies remains fragmented; with financial institutions, namely MSCI and Sustainalytics, attempting to gather tangible data to issue ESG scores.

Non-financial Reporting Will Become a Prerequisite

Writing about ESG and non-financial reporting standards is no longer enough. It is all about meaningful non-financial reporting. This includes figures quantitatively with integrated granularity on the company's plans to accomplish. It is an area on which the IRO, management and the board of directors should focus significantly. Financial reporting remains the most important information source for an investor.

The volatility in the financial markets reflects all information flow and future expectations in the market. Therefore, a superb financial report is today not enough to satisfy the financial community. Equity analysts will use alternative sources to find an information advantage. Investors will use this knowledge combined with other sources and discussions with management to contemplate the financial report.

Likewise, superb non-financial reporting is also not enough to satisfy the financial community or the company's stakeholders. Today, institutional investors are increasingly using their ESG specialists or external ESG research

P. Lykkesfeldt and L. L. Kjaergaard, *Investor Relations and ESG Reporting in a Regulatory Perspective*, https://doi.org/10.1007/978-3-031-05800-4_39

companies to conduct ESG research on both listed and unlisted companies. In addition, rating agencies have sprung up to compare companies and serve as useful information tools to the financial community on which companies are best-in-class and lagging.

Aside from delivering a professional non-financial report, this will merely be a prerequisite for any listed company in the future. The reporting must be integrated with the equity story and stakeholder considerations to attract and maintain institutional capital.

Overview of the ESG Rating Agencies

Financial data is by no means perfect as a company's outlook, guidance and strategy remain self-reported. In addition, company earnings are affected by accruals, and R&D only measures cost input and not beneficial output. Yet, most financial data is relatively objective, independent and transparent.

However, due to the lack of validated standards, the market for gathering ESG and non-financial reporting data is highly unconsolidated. Today, over 600 ESG rating agencies[1] deploy varying methodologies, focus areas and processes. In addition, the financial community attempts to cultivate their frameworks and implement their ideas—the outcome result is less transparency and more subjectively. As a result, two topics are out of the company's hands, which it must mitigate (Fig. 39.1).

Twenty-eight out of the world's largest institutional investors used four or more ESG rating agencies in 2021. Conversely, only twelve of the world's 50

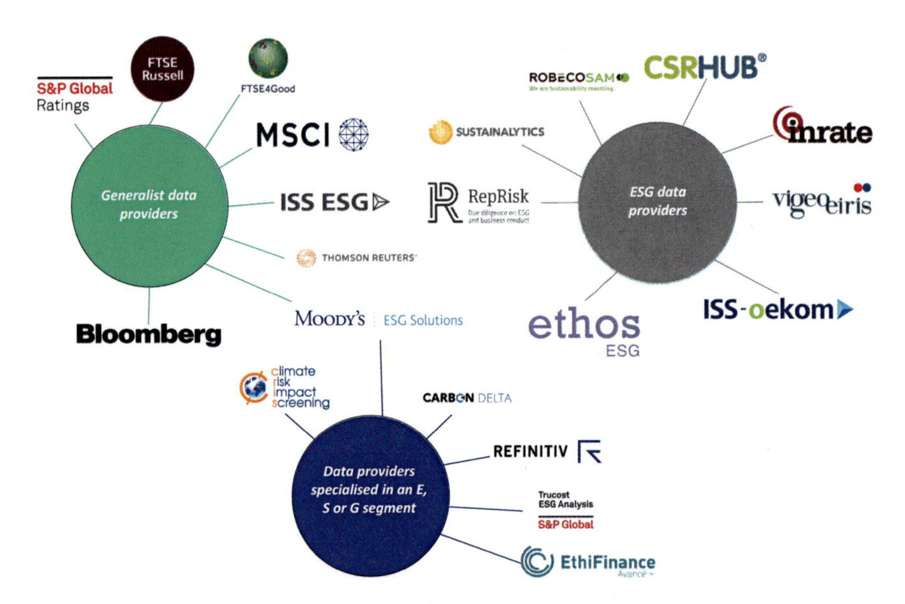

Fig. 39.1 Landscape of ESG agencies[2] (*Source* SIA Partners [2020])

largest investors used one or no agencies.[3] This is opposed to gathering financial data. Most investors and equity analysts only use one or two platforms, namely Bloomberg, Refinitiv Eikon, FactSet or Capital IQ, that together have ~64% market share.[4]

We believe that the key reason is the lack of standards, which we believe taxonomy, CSRD and SFDR can assist within Europe. It is the duty of the IRO to be updated on the sources of ESG data and assist the financial markets with aligning its frameworks to how the company sees it. The EU will likely create a common European ESG database to increase transparency and democratisation of data.

The Significant ESG Rating Agencies

Sustainalytics (acquired by MorningStar in 2020) and MSCI currently serve as the world's most influential ESG rating agencies. They make the data publicly available; that is, they offer subscription options to access and analyse the data easily. In general, the ESG rating agencies' rating of companies is based on a risk assessment of their non-financial reporting materiality, measurable risk and engagement (Fig. 39.2).

There has been noteworthy debate and discussion in the past years about the differences in the rating agencies. In the past, it seemed that a large proportion of these focused on onboarding the (the quite untransparent) ESG data on as many companies as possible. This would equate to dominance and consequently a more viable business model. However, as companies increase transparency and the rating agencies become more professional, so does the quality of the data. In terms of the ESG rating agency, we find it fair to expect both a significant growth and consolidation as this industry will mature over time.

Despite the increased professionalisation, attention has been called for analysing the underlying data and disconnecting the rating itself from how the data is presented.[6] The divergence of the ratings can be attributed to three factors: scope, measurement and weighting. Research has found that the factors contributed 38, 56 and 6% of the divergence, respectively:

- **Scope**: How many variables a rating agency applies can significantly impact its rating. For example, a company may score exceptionally high on some ESG metrics and low on others. Including or not including the best and worst-performing scores will impact the average of the scores.
- **Measurement**: If rating agencies apply a similar scope, they may diverge in which indicators are applied. There are especially high variations for corporate governance, as some agencies may consider some unethical behaviour environmental or social, and others consider them a breakdown of governance policies.
- **Weight**: When generating an ESG rating, the agency will weigh the scope and measurement data relative to their preferences. Sometimes these

1. Materiality assessment

Material assessment principles from a regulatory point of view as well as analysis from the provider's database

An initial assessment of standard governance issues as well as a selection of 3-10 relevant ESG issues

2. Scoring the Exposure risk/Management risk

Issue exposure x Beta − Issue managed risk = Issue unmanaged risk

3. Final ESG rating & company's engagement

Each issue's score is normalised to the industry of the issuer, taken out of the last three years of all companies rating, and the unmanaged risk is converted into a score from 0 or CCC (worst) to 100 or AAA (best)

Fig. 39.2 ESG as a rating tool[5] (*Source* Briand, R. [2021])

weights will change depending on the rating. However, statistical analysis shows a small divergence in this factor (Fig. 39.3).

The Transformation of the Equity Analyst

Financial and non-financial reports are generally backward-looking, whereas the financial community is forward-looking. In other industries, notably law, divergences of information are condemned. However, some see the divergence as added value and beneficial to pricing assets in the financial community.[8] Of course, the data must be as factual as possible; however, the underlying framework of weights and intangible aspects of non-financial reporting is subjective, just like the DCF assumptions.

The IRO should respect this and work with agencies on the validity of the data itself, but not challenge the weighting. The IRO is used to working with equity analysts (Chapter 11) and may manage expectations and nudge the agencies to consider various options. However, the IRO cannot keep track of 600 rating agencies, so it must select the most relevant and understand their

Categories	Sustainalytics (Morningstar)	MSCI	ISS OEKOM	Moody's ESG solutions	S&P ESG	Refinitiv (Thomson Reuters)	FTSE Russell	RepRisk
Team size	200+	270+	180+	120+	100+	150+	n/a	100+
Coverage	12,000+	14,000+	6,000+	5,000+	7,000+	9,000+	7,000+	170,000+ (incl. private companies)
Score (Min. to Max.)	0 to 100	CCC to AAA	-D to A+	0 to 100	0 to 100	0 to 100	0 to 5	1 to 100
Rating cycle	Annual	Annual	Annual	Annual	Annual	Weekly	Annual	Daily (AI-based)
Subscription needed (Y/N)	No	No	Yes	Yes	No	No	No	Yes
Indices supplied	Solactive, Stoxx, S&P	MSCI, Bloomberg	STOXX, Solactive	EuroNext	S&P, DJSI	Refinitiv	FTSE	FTSE, DowJones, S&P
Data verification by companies	Yes	Yes	Yes	Yes	Yes	Yes	Yes	No
Second party opinion	Yes	Yes	Yes	Yes	Yes	No	No	No

Fig. 39.3 Overview of the largest rating agencies[7] (*Source* Huck-Wettstein, M. [2020])

methodologies and data. This is especially relevant for passive investments, as they track the data on an aggregated basis.

Moreover, as the company is forecasting on financial guidance, forecasting on non-financial guidance must be strongly considered. Balancing competitive concerns with forecasting is just as material for non-financial reporting as financial reporting.

Given the extreme current and expected future demand for ESG, embracing divergence is important for a company. It is especially evident considering the high-cost effects on investment banks from MiFID II's unbundling of research and trading commissions (Chapter 14). Moreover, based on sources from Frost Consulting, UBS and the Manager Sustainability Reports, the Financial Times (18 January 2022)[9] mentioned the development of broker costs and projected that ESG is a major opportunity for them. Therefore, divergence and subjective options on non-financial reporting will stay; however, the baseline of data will likely become professionalised. To continue a strong relationship with institutional investors, we believe the equity analysts will increase their focus on ESG, which the IRO may benefit from (Fig. 39.4).

CSRD Will Professionalise Baseline Data

There is an established market for rating agencies within ESG, which conduct independent studies on the company's ESG figures. CSRD will likely complement these independent studies, allow greater transparency and increase investor confidence in the published data. The divergence between non-financial reports and the rating agencies will equate to less confidence in a particular company.

A company has the best available resources to predict its long-term cash flow and justify its intrinsic value (Chapter 2). It should then be a main

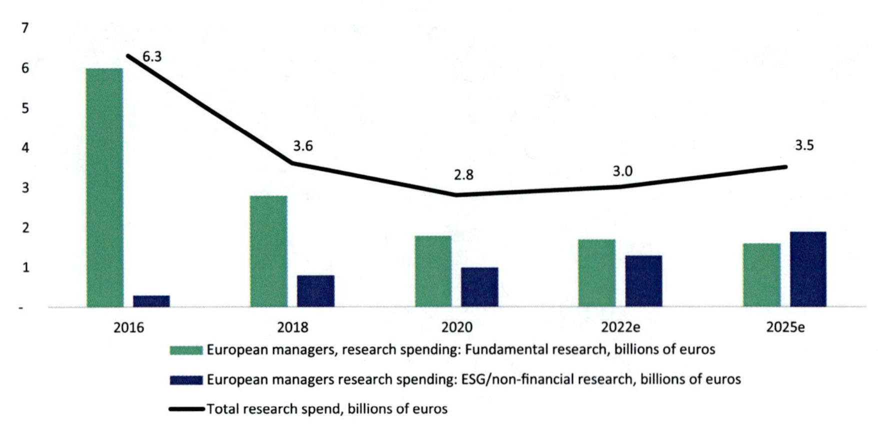

Fig. 39.4 Research spending on sell-side research (billions of euros) (*Source* Klasa, A. [2022])

objective for the IRO to mitigate the company's risk premium. Likewise, a company has the best available resources to judge its non-financial impact. It should, therefore, also mitigate the difference between its non-financial impact and the ESG data providers to lower differences. Just like understanding and respecting the importance and the role of equity analysts and brokers, ESG rating serves a similar purpose. They increase information flow so that many stakeholders receive a transparent news flow on non-financial matters.

As non-financial CSRD-compliant data become available, we believe there will be noticeable differences to the ESG rating agencies. As the company sets its non-financial standards, it will have to watch the rating agencies. The company must discuss and encompass the reporting lines with the ESG agencies to ensure alignment.

A summary of facts and best practices is illustrated on page 329.

References and Sources

1. ERM. (2020). *Rate the raters 2020.* The sustainability institute.
2. SIAPartners. (2020). The ESG data market: Changes and challenges for financial service providers. https://www.sia-partners.com/en/news-and-publications/from-our-experts/esg-data-market-changes-and-challenges-financial-services.
3. Brady, A., et. al. (2021). *Managing ESG data and rating risk.* Harvard Law School Forum on Corporate Governance.
4. Burton Taylor. (2020). *Financial market data report.* BT International Consultation.
5. Briand, R. (2021). *Ask your shareholders which ESG yardstick to use.* Nordea.

6. Berg., F. et. al. (2019). Aggregate confusion: The divergence of ESG ratings. *SSRN:* https://ssrn.com/abstract=3438533.
7. Huck-Wettstein, M. (2020). ESG ratings and ranking: Why they matter and how to get started. *Sustainserv:* https://sustainserv.com/en/insights/esg-ratings-and-rankings-why-they-matter-and-how-to-get-started/.
8. Hirai, A. (2021). Managing ESG data and rating risk. *Harvard Law School Forum on Corporate Governance.*
9. Klasa, A. (2022). Sustainable funds face threat from tech sector. *Financial Times.*

PART X

Future Trends of Financial and Non-financial Reporting

CHAPTER 40

Best-Practice IR is About Being at the Forefront

Prediction is very difficult, especially if it's about the future.

—Niels Bohr, Physicist

Integrating Consumer and Investor Protection

In this book, we have focused on the strategic implementation of best-practice IR and best-in-class non-financial reporting. We have discussed the origin of ESG and raising the bar to stakeholder capitalism. The UNGC has focused on top-down strategic implementation of the SDGs. Major institutional investors embrace responsible investing, and the EU is accelerating the push to integrate people, prosperity and the planet adjacent to the profit-only mentality.

The world is lagging behind the SDGs, the Paris climate agreement's goal of carbon neutrality by 2050 remains ambitious and greenwashing, and window dressing remains challenging to combat. Compared to the Paris climate agreement's "well below" 2 degrees Celsius target by 2030, the climate action tracker currently estimates 2.7 degrees Celsius with current policies. If the target of 55% emission reduction by 2030 is achieved, then 2.4 degrees Celsius is more realistic and optimistically 2.1 degrees if all submitted targets are met.[1] This is a major challenge for the world that requires action from all stakeholders.

We wish to underpin the purpose of considering stakeholder capitalism to understand the next generation of EU policy—and therefore, by the IRO, senior management and board of directors can further optimise its function in the future legal framework. The existing legal framework in the financial markets aim to protect investors by increasing transparency and limiting

P. Lykkesfeldt and L. L. Kjaergaard, *Investor Relations and ESG Reporting in a Regulatory Perspective*, https://doi.org/10.1007/978-3-031-05800-4_40

market abuse. A key focus of MiFID/R's has been the classifications of the types of investors. Financial institutions need to be more aware and adjust accordingly to the protection level corresponding to the knowledge of different client groups, and henceforth also consider their sustainable preferences in their investment strategies. This is particularly relevant in respect of retail investors.

In parallel, the EU is currently introducing new legislations and standards on the safety and use of consumer goods, including product safety and trade policy on goods. Perhaps not from a legal perspective, but the legislations and mentality of the two branches will become more intertwined. The EU's New Consumer Agenda of 2020 and the associated legislation[2] have outlined that the aim is to empower European consumers to make informed decisions and play an active role in the ecological and digital transition. In 2022, the EU commission will likely also introduce revisions to the "sales of good directive" on encouraging more sustainable circular products.

There is a clear need to register and track progress, as talking about green initiatives and objectives is no longer enough. In a new study among board members,[3] 75% said that climate change is important to strategic success, yet 43% said their company has no carbon targets, and only 16% have scope 2 and 3 targets. Also, 16% said that no one in the organisation is responsible for non-financial reporting. Due to the complexity of non-financial reporting and ESG, the individuals with expert knowledge on the topic and its implications need to have a strong collaboration with senior management and the board members who are responsible for business implementation. That is naturally a significant alignment challenge.

Consequently, the EU is pushing for the investment community, companies and consumers to consider their environmental impact. To do this, transparency is needed to know the exact characteristics and use of the products (financial and non-financial) that are being sold and traded. The climate agenda must be included in board rooms, and the ESG responsibilities addressed in this book must be anchored to specified individuals.

Considering Consumer Activism

Consumer theory on choice, behaviour and trends are critical pillars in microeconomics. Currently, we witness several key trends of online shopping, anonymity, transparency and demand for clean and green products. Due to the rise of digitalisation, organised boycotts are getting more common. Companies are forced to embrace the bottom-up mentality that the organised consumer is an important platform collectively. Therefore, many companies also embrace democratisation, so consumers have more personal contact with the company's representatives or even the CEO.

Retail investors own 12% of total financial assets, and most of the institutional capital is funded through pensions and taxes.[4] Therefore, we may find that the bottom-up mentality already witnessed for consumers can spread to retail investors. Utilising digital IR and FinTech solutions can allow the IRO to service retail investors on par with the servicing of institutional investors. Likewise, some companies—notably Spotify—conducted their IPO without the involvement of an investment bank or contact lists to institutional investors.

With the EU as a cornerstone for raising the bar for institutional investors, all fiduciaries are required to act in the best interest of their clients. Moreover, with increased transparency and democratisation, the clients can improve their pressure and engagement towards the institutions. This will result in increased competition among institutional investors, who will, in turn, also increase pressure on companies to deliver on their ESG pledges.

The Digitalisation of Non-financial Reporting

Some companies choose to outsource some of their direct IR responsibilities, most notably in the US and the UK, although this is not so common in Continental Europe. However, outsourcing of non-financial reporting may well grow in the coming years as more companies are required to report on their ESG statistics, and due to the complexity of the area. To ensure comparability, benchmarking and transparency, digital solutions are currently being developed. Companies that specialise in outsourced IR and non-financial reporting benefit greatly from the economics of scale, compared to each company (especially SME), creating a fully consolidated process.

WEF has stated that 70% of the underlying SDGs will be solved by harnessing the power of new technologies in the fourth industrial revolution.[5] In addition, ESMA has included “digital tagging for a machine-readable format” in their taxonomy guidelines. Therefore, it is evident that artificial intelligence will become an essential aspect of integrating non-financial reporting.[6]

In addition, the complexity and the level of quality and detail will increase gradually for companies. A key reason is that increasingly companies are utilising blockchain for the effective supply chain management. The ability to track every good and underlying component is a great advantage for companies and is easily compatible with integrated reporting.

European Single Access Point (ESAP)

In November 2021, the European Commission published proposals for an ESAP.[7] ESAP will be a centralised database for the financial community and stakeholders to view the reporting and compliance of companies in **the** EU. Considering the regulatory wave and the new ESG framework, companies must disclose a wide range of information. It will likely be phased in between 2024 and 2026.

ESAP will contain information on MiFID/R, MAR, SFDR and taxonomy legislation. The centralised database should be a helpful resource to increase transparency and ease of communication. Systems and controls will have to be updated to ensure that companies and investment firms submit and correct information at correct intervals. Information on the timeline and implication remains at early stages.

MiFID III

Alongside the ESAP, the EU also, in August and November 2021, announced coming amendments to MiFID II/MiFIR. This was highly anticipated—and dreaded—by the financial community, particularly investment banks. However, the proposals are limited in nature and are largely aligning amendments to the already published SFDR.

The main attribute of MiFID II was trading reporting to increase transparency on trading fees and to increase the protection of investors. MiFID III will likely include establishing a centralised database for market data and technical considerations on harmonising data reports for trading financial instruments. In addition, we believe that MiFID III may include amendments on how to disclose reports on best execution, non-equity translations and the provision of data. In general, we find that these themes will have limited influence on the work of the IRO.

Still Early days for the US' Considerations on Non-financial/ESG Reporting

The EU legislators are at the forefront of implementing ESG policy through sustainable investments, the taxonomy framework, CSRD and SFDR. The UK has also followed suit with its Sustainable Disclosure Requirements (SDR). However, the corresponding regulation in the US is under development as legislators have not yet mobilised to reduce climate risks. Yet, it seems that the landscape is changing (highly correlated with the current US President's priorities). On his first day in office, US President Joe Biden re-joined the Paris agreement. Moreover, in May 2021, he signed an Executive Order on Climate-related Financial Risks designed to mobilise legislators to accelerate their efforts on climate risks.[8]

Despite the proactive actions from the US's government, the legislation at the state level is more varied, including in Texas that recently passed legislation preventing state pension funds from boycotting its financial positions in the oil and gas industry.[9]

As a next step, US President Biden asked the Financial Stability Oversight Council (FSOC) to publish a report with proposed legislation to mitigate climate-related financial risks to the financial system and assets. The FSOC is currently chaired by the US Treasury Secretary, Janet Yellen, the chairs of the Federal Service, Jay Powell; of the Securities and Exchange Commission, Gary Gensler; and the other distinguished members. The FSOC published its report on 21 October 2021.[10] The report provides recommendations on:

- **The assessment of climate-related financial risks** to financial stability, including through scenario analysis, and evaluate the need for new or revised regulations or supervisory guidance to account for climate-related financial risks.
- **The enhancement of climate-related disclosures** to give investors and market participants the information they need to make informed decisions, which will also help regulators and financial institutions assess and manage climate-related risks.
- **The enhancement of actionable climate-related data** to allow better risk measurement by regulators and in the private sector and
- **Building capacity and expertise** to ensure climate-related financial risks are identified and managed. This includes recommendations for the Securities and Exchange Commission (SEC), the Federal Reserve Board (FSB), the Commodity Futures Trading Commission (OTC) and the Federal Housing Financing Agency (FHFA).

Already in March 2021, the SEC created a Climate and ESG task force to address greenwashing and address potentially false and misleading ESG-labelled products to increase investor protection and ensure market transparency.[11] In addition, considering the FSOC report, the SEC is following the recommendation to set a reporting standard for non-financial ESG data—in line with the EU's regulation. These efforts should ensure mandatory climate disclosures and be the first step to become a baseline for investment and AGM voting decisions.

In March 2021, the FSB founded a Supervision Climate Committee to determine the central bank's credit obligations risk relative to different climate scenarios.[12] In addition, it launched a Financial Stability Climate Committee to navigate the recommendations from the FSOC.

In December 2021, US President Biden also signed an executive order to dramatically reduce CO_2 by aiming for net-zero emissions in all federal options by 2050, and a 65% reduction by 2030.[13] With the US President Biden administration, FSOC, SEC and FSB, it seems that the US is moving ahead with an ESG agenda; however, it is still early days in terms of the implementation of transparency and disclosure measures. As the ISSB is currently developing its harmonised guidelines for non-financial/ESG reporting, it seems that the currently under-developed legislation in the US will follow the ISSB, once established.

References and Sources

1. CAT. (2021). Glasgow's one degree 2030 credibility gap: Net zero's lip service to climate action. CTA: https://climateactiontracker.org/documents/997/CAT_2021-11-09_Briefing_Global-Update_Glasgow2030CredibilityGap.pdf.
2. EU. (2021). Regulation of the European parliament and of the council on general product safety, 2021/0170(COD). *European Commission*.

3. Besland, L. et al. (2021). Changing the climate in the boardroom. *Heidrick & Struggles:* https://www.heidrick.com/en/insights/sustainability/changing-the-climate-in-the-boardroom.
4. Rickenbacher, R. (2022). Private investors are crucial to the future of ESG: Here's how they can find their voice. *Blog of the World Economic Forum.*
5. WEF. (2020). Unlocking technology for the global goals. *World Economic Forum.*
6. Macpherson, M., & Schmid, S. (2021). Weathering the storm of ESG complexity by leveraging AI. *Oddo bhf: Sustainable Finance.*
7. EU. (2021). Regulation of the European parliament and of the council on establishing a European single access point providing centralized access to publicly available information of relevance to financial services, capital markets and sustainability. 2021/0378(COD). *European Commission.*
8. Biden, J. (2021). Executive order on Climate-related Financial risks. *Presidential actions:* https://www.whitehouse.gov/briefing-room/presidential-actions/2021/05/20/executive-order-on-.
9. Texas senate bill. (2021). Related to state contracts with and investments in certain companies that boycott energy companies. *TX state legislator:* https://legiscan.com/TX/text/SB13/id/2332616.
10. FSOC. (2021). Report on climate-related financial risks. Financial stability oversight council. https://home.treasury.gov/system/files/261/FSOC-Climate-Report.pdf.
11. SEC. (2021). SEC announces enforcement task force focused on climate and ESG issues. *Press release:* https://www.sec.gov/news/press-release/2021-42.
12. Schroeder, P. (2021). Fed privately presses big banks on risks from climate change. *CNBC.*
13. Biden, J. (2021). Executive order on catalyzing America's clean energy economy through federal sustainability. *Briefing room:* https://www.whitehouse.gov/briefing-room/statements-releases/2021/12/08/fact-sheet-president-biden-signs-executive-order-catalyzing-americas-clean-energy-economy-through-federal-sustainability/.

CHAPTER 41

Is the Next Decade Scary or Exciting?

Moving Beyond GDP

Stakeholders, the investment community, policymakers and companies are increasingly looking at ESG factors to compliment the profit-driven motive. The criticism of capitalism and the profit-driven agenda is not new. Without moving into a political and sociological debate on the development of economics, it is important to note that capitalism's fundamental and useful critique has been championed by Karl Marx, David Ricardo and more recently, Thomas Piketty. Yet, capitalism and the financial markets have been tremendously adaptive and flexible to societal and technological advances for the past century.[1]

We seem to transition into a new type of capitalism, melding the creation of prosperity, serving people and society and caring for the planet. The development of ESGs seems to be closely tied with the debate on the gross domestic product (GDP) indicator. GDP measures the monetary value of all goods and services produced and sold by a country and is a useful—yet conventional—measure of a country's performance.

As earlier outlined, environmentalism and social justice became largely mainstream in the 1970s/1980s (Chapter 30). Two months before his assassination, Robert Kennedy famously gave his first presidential campaign speeches on the anti-Vietnam war protests. In the speeches, he remarked that a country's GDP is an insufficient measure of success. The reason is that it ignores societal (e.g. real GDP per capita and median incomes) and environmental factors (e.g. "the ecological deficit" of the planet). This debate has reascended in recent years.

ESG metrics for a sustainable future and the "beyond GDP initiative" were key speaking points at the World Economic Forum's (WEF) Davos agenda

P. Lykkesfeldt and L. L. Kjaergaard, *Investor Relations and ESG Reporting in a Regulatory Perspective*, https://doi.org/10.1007/978-3-031-05800-4_41

in January 2022. The EU, OECD and WWF all sponsor the beyond GDP initiative. The initiative is developing indicators as clear and appealing as GDP but more inclusive of environmental and social aspects of progress.

WEF members assert that instead of having GDP as the only factor, policymakers, investors and companies should consider additional factors in their decision-making. This means adding three additional three Ps, alongside profit: people, prosperity and the planet. This is like ESG, and we believe these could be translated into stakeholders, purpose (as advocated by CEO Larry Fink of BlackRock) and environmental factors for the financial community.

Profit, People, Prosperity and the Planet

In investigating non-financial and ESG reporting, it becomes clear that the fundamental benefits and drawbacks of the financial markets remain unchanged, so does the factors determining share prices and how to communicate to the financial community. However, adopting ESG in equity analysis is rising rapidly.

Including non-financial factors is becoming a key pillar in valuing a company's positioning and opportunities in competitive markets. Many stakeholders outside the world of investing believe that the financial markets are only driven by generating profits. This is partly true. However, professional investors are more deeply concerned about the equity analysis of how the company generates profits. We believe that the financial community is already embracing these factors, and they are already accelerating into the decade for action.

Moving into a Decade of Action

Given the current inadequate progress made on the SDGs, the UNGC has coined the 2020s the "decade of action" and the "decade of delivery". There is an interest and need to include people, prosperity and the planet in all investment and strategic decisions. In addition, the EU has stated that the green deal requires "digitalisation as an enabler for decarbonisation" and has "technologies that work for people".

As a result, it is clearer that the past 100 years of a profit-only agenda will be less applicable for the next decade. In the past five years, we have witnessed most countries joining the Paris agreement of limiting global warming, the SDGs and now the new wave of the ESG regulatory framework.

We are moving beyond profit and growth to now considering the company's purpose, its stakeholders and the planet's well-being. This will incur several megatrends, moving from digital to the decarbonisation sector. This is unmistakable as today we begin to question the use of technologies, considering not if they are possible to implement, but why and what we need from them. As part of the EU's "six thematic clusters" of megatrends (Chapter 16), companies and investors (cluster 5) need to consider the movements of people (cluster 1) and include the accelerating technological changes (cluster 4). We have listed a few examples of food, energy and lifestyles below.

Cluster 2: Climate Change and Environmental Degradation Worldwide

- A worldwide study among 10,000 young people (aged 16–25) on climate change and anxiety found that nearly 60% of the people are extremely worried or very worried about climate change.[2]
- The EU has already announced carbon taxes on flying and meat products. Introducing regulatory "nudging" can ultimately result in decarbonisation on an individual level.
- Today, 75% of the world's energy is derived from oil, gas and coal. Solutions including wind, solar, hydro, next-generation nuclear and perhaps fusion power have clear opportunities. Especially on the African continent, the figure is as high as 97%, cited by the EU Commission in the 2022 Davos Summit.
- In addition, there is an inadequate supply of industrial and reactive metals for the quest of electrification to reduce emissions. In the extraction of the metals, the grade is diminished, meaning that relative water and electricity consumption increase. Alongside the environmental implications, the social implications are that most of these metals are found outside the Western world.

Cluster 3: The Increasing Scarcity of Global Competition

- 30% of CO_2 emissions are from agriculture and livestock, and there is a debate if this can keep up with population and wealth growth.
- Solutions exist, including vertical farming and plant-based and lab-grown meats. However, the solutions must be adaptable, sustainable healthy and affordable.
- The WEF has predicted that a "nature positive economic" can generate ~9 trillion euro in annual business opportunities and 400 million jobs by 2030.

Cluster 6: Diversifying Values, Lifestyles and Governance Approaches

- Following the COVID-19 crisis, the buzzword has been "the great resignation" as many people have decided to resign from their jobs and rethink their lifestyles. In addition, work automation will likely diminish with integrating IT solutions such as machine learning and artificial intelligence. As a result, employees will work less but more effectively, focusing on "people". As the EU has proclaimed, this will result in working-from-home and virtual working and balancing human values.

- In education, schools already include digitalisation in classrooms. As a result, students will likely move from receiving and processing information (as the acceleration is exponential) into becoming life-long learners on learning effectively.
- We hope to focus on consumer spending and decarbonisation, pledging on people to track their CO_2 footprint. Tracking emission footprint is already possible on some bank and credit card platforms, and increasingly, we witness fashion and retail companies introducing "CO_2 labelling" on the packaging.
- Data will be introduced into the healthcare industry. This included personalised medicine and combining food and lifestyles with health care. Introducing data will equate to more affordable, personalised and connected solutions for the industry.

Considering the Role of SDGs in Corporate Financing

Moving into the "decade of action" (as referred to by the United Nations Global Compact) is not only essential for all participants of the equity markets, but also for the debt and credit markets. The budgets to achieve the SDGs on the 2030 agenda remain short by 2.2 trillion euros (2.5 trillion US dollars) annually.[3] As CFOs are the stewards of trillions of investments, the UNGC in 2019 initiated the CFO network to mainstream the SDGs on the capital markets. Mobilising capital, issuing SDG-linked bonds and including SDGs in foreign direct investments will assist in closing the budget gap to achieve the 2030 agenda.

To continue the critical work, the CFO network introduced and published its principles in 2020.[4] The four principles are: (1) SDG impact and measurement, (2) integrated SDG strategy and investments, (3) integrated corporate SDG finance and (4) integrated SDG communication and reporting. The interest in ESG-linked bonds is growing steadily, yet it remains a contemporary and young topic for the financial community. For example, in 2020, one of the largest pharmaceutical companies in the world, Novartis, launched a 1.85 billion euro (2.1 billion US dollar) ESG-bond tied to access to medicine. Instead of a profit-driven goal, the incentive of the bond was to incentivise the company to broaden its reach to more patients. Yet, the bond issue has been criticised for the perceived confusing metrics, little transparency and blurred lines between intentions and meaningful action.[5]

It is yet not clear how non-financial/ESG reporting can completely be integrated, and these limitations are also evident on the bond market. As this book focuses on the equity markets, and as the themes of the debt and financing markets are gaining traction, we find that this area should benefit from coming independent studies.

The Next Decade Will Be Beneficial for Those Who Embrace It

In 2022, we will understand more about WEF's ISSB, to help consolidate non-financial reporting given the current lack of comparability, reliability and relevance. Yet, non-financial reporting remains difficult. We hope that the technical working group will consider implementation easiness high. As companies focus on the massive business opportunities in digitalisation and decarbonisation, we believe it is vital to remember the key strategic objectives. Alongside profit, these objectives need to encompass people, prosperity and the planet on the organisation's strategic, operational and stakeholder level.

The financial community are increasingly opening their mind to stakeholder capitalism, and we believe the IRO must navigate the company's risk premium. The risk premium constitutes financial reporting and increasingly will encompass non-financial and ESG reporting. Therefore, the IRO must utilise two-way communication with the financial community to shape the ESG communication and shape the organisation in a strategic role.

Legislation on the financial markets aims to ensure and promote fair and transparent markets. Instead of viewing MiFID/R, EMIR, MAR and taxonomy as a bump on the road, we believe organisations—headed by the senior management and board should embrace the underlying themes and integrate them with their business. To end our book, we would like to highlight three key questions posed in a recent article by the former CEO and Director of UNGC, Lise Kingo, to the board meetings in 2022[6]:

1. Is ESG fully integrated into our overall business strategy and cascaded across the organisation into transparent integrated reporting?
2. Do we have a financed and operational net-zero climate strategy aligned with the Paris Agreement and 1.5 degrees C ambition for Scope 1, 2 and 3?
3. Are we ready for the new company reporting requirements on non-financial reporting?

The companies who can execute adequately will also find the best business opportunities for their financial statements whenever they need to communicate or utilise the benefits of the financial markets.

A summary of facts and best practices is illustrated on page 329.

References and Sources

1. Acemoglu, D., & Robinson, J. (2015). The risk and decline of general laws of capitalism. *Journal of Economic Perspective, 29*(1), 3–28.
2. Marks, E., et al. (2021). Young people's voice on climate anxiety, government betrayal and moral injury. *SSRN*. https://ssrn.com/abstract=3918955.
3. UN. (2020). The sustainable development goals report 2020. *United Nations*.
4. UNGC. (2020). Introducing CFO principles on integrated SDG investments and finance. *CFO taskforce for the sdgs*.
5. Plüss, J. (2022). Can Novartis' sustainability-linked bonds make good on its promises? *Swissinfo*.
6. Kingo, L. (2021). 3 simple questions for your board going into 2022. *Greenbiz*. https://www.greenbiz.com/article/3-simple-climate-questions-your-board-going-2022.

CHAPTER 42

Postscript

Companies, investors, NGOs, governments and other stakeholders are scrambling to include ESG considerations in their processes. Yet, when researching for this book, we felt that most institutions focus on the "how" and not on the "why" for doing so. We, therefore, find it appropriate, again, to highlight the UN's considerations on not only thinking about profits, but also prosperity, people and the planet in their practical approach and descriptions to integrating non-financial/ESG reporting.

An institution should always include its purpose when managing its approach and actions, otherwise what is the meaning?

The UN has now lightened the torch for a sustainable world economy, and with its new significant legislative steps, the EU is now carrying the torch. Hence, the question arises, who will carry it next? By re-joining the Paris agreement, US President Biden took his first steps by signing executive orders on mitigating climate-related financial risks. Yet, the US government institutions seem to be on hold to take the ambitious steps that the EU has taken. The financial and stakeholder community are all awaiting the highly anticipated disclosure standard by the ISSB's technical readiness working group, which we believe will be helpful for the further progress.

While we wait, only five of the 17 SDGs are "on track", and six are in complete "deterioration"—and even in the most optimistic scientific scenarios, the world will overshoot the science-based targets of maintaining the global average temperature "well below" 2 degrees Celsius versus the pre-industrial levels.

In this book, we discuss the term "you get what you measure" in that institutions must incentivise their senior managers and employees both financially

P. Lykkesfeldt and L. L. Kjaergaard, *Investor Relations and ESG Reporting in a Regulatory Perspective*, https://doi.org/10.1007/978-3-031-05800-4_42

and non-financially to embrace their ESG processes—just like they do for the financial performance.

Yet, for all the current initiatives and inclusion of ESG in their processes currently carried out by various institutions, we need all governments on a global scale, not only the EU and the US, via relevant legislation, to get involved and embrace non-financial reporting/ESG, with a view to consolidate non-financial reporting/ESG on a global scale.

On this background, our best, but still humble, conclusion is:

> "If you can't measure it, you can't improve it".

CHAPTER 43

Summary of Facts and Best Practice

Part I: The Financial Markets: An Overview

- The financial market is the world's largest workplace, worth 370 trillion euro (422 trillion US dollar).
- Positive aspects of a share listing:
  - Easier access to capital.
  - Liquidity in the ownership.
  - Transparency and servicing minority shareholders and stakeholders.
  - Ease of ownership and exit route to investors.
  - Creditability and marketing.
  - Employee engagement.
- Negative aspects of a share listing:
  - Time consumption for management.
  - Likelihood of takeover risk increases.
  - The direct cost of being listed.
  - Increased transparency and lead to lower competitive advantage.
  - Costs of exercising best-practice IR.
  - Risk of short-termism from equity holding managers.
- The aggregated assumptions of investors drive the share price. As a result, understanding the flow of information is essential to know how to operate in the market.
- Beware of the company's risk premium, the difference in full fair value of company and its share price.
- Controlling the risk premium leads to improved return to shareholders, reduced cost of capital, and reduced risk of takeover.

P. Lykkesfeldt and L. L. Kjaergaard, *Investor Relations and ESG Reporting in a Regulatory Perspective*, https://doi.org/10.1007/978-3-031-05800-4_43

- The risk premium can be influenced by a best-in-class IR function and higher transparency, which leads to less uncertainty.
- Beware of types of information, investors and investment strategies—as well as the fundamental assumptions of the DCF.
- ESG is 10% of total invested capital and there is established academic research on the benefits of responsible investing and integrating ESG in the DCF framework.

Part II: The Participants of the Financial Markets

- Have respect for equity analysts and treat properly, even if they are negative in their company recommendations.
- Work for good relationships but keep a professional distance.
- Be in control—do not be controlled by the equity analysts or investors.
- Incorporate an appropriate custom for comments on the equity analyst's reports in draft form. Point out factual errors in equity analysts' reports when these misunderstandings can be verified elsewhere.
- Never enter confidential relationships with an equity analyst or investor.
- Do not give in to the equity analyst's eternal desire to see management if the company has a satisfactory IR function. However, equity analysts should occasionally meet with the company's management.
- Be careful in choosing an equity analyst and broker when preparing for roadshows and investor meetings.
- Get references regarding the corporate finance adviser from the management of other companies.
- Do not get dependent on one investment bank.
- Always seek to allow at least two corporate finance advisers to compete against each other in the selection phase.
- Evaluate and select a corporate finance adviser according to a structured procedure, including the evaluation tool mentioned.
- A corporate finance advisor's proposed fee can always be negotiated.
- After all, they have a healthy critical sense of corporate finance advisers' assessments and recommendations. They are, after all, very transaction-oriented due to the typical no cure-no-pay fee structure.
- Have respect for investors—also the smaller ones and the retail shareholders.
- Work for good relationships with investors but keep a professional distance.
- Be consistent in the company's announcements.
- Never enter confidential relationships with an investor—even if they own a significant shareholding in the company. Most professionals, however, refrain from coming close to confidential knowledge.
- Use investors' thoughts and ideas constructively.
- Investment banks will need to prove the quality to get investors to pay.
- Quality requires good data, analytics and expertise.

Part III: Major Legislation Themes Related to the European Financial Markets

- Companies must have formal best-practices policies and procedures in place for MAR. This includes the identification, mapping, monitoring and reporting of suspicious activity and market abuse.
- Staff must receive regular compliance training to ensure that compliance and IR procedures and policies are being implemented effectively.
- All legislation on the financial markets aims to ensure and promote fair and transparent markets.
- Be aware of MiFID II's considerations on inducements, investor classification and conflict of interest.
- Staying compliant with MAR is twofold. On the one hand, entities need to disclose and record all inside information compliant promptly. On the other hand, the company must ensure processes to identify, report and record abuse to the relevant authorities.
- EMSA's guidelines and associated technical standards on disclosing and processing insider information must be complaints. Up-to-date insider lists, standards and protocols must be updated and communicated to the staff of its importance.
- Any suspicious activity should be reported directly to the relevant authorities, accompanied by the relevant audit trail to support due diligence. In addition, entities should keep precise records for up to five years, using these records and alerts as a route to reduce risk and improve the monitoring system's accuracy.
- Investment banks are under pressure from MiFID II's considerations on unbundling fees and requirements on corporate governance.
- The IRO should consider commission research, commissioned corporate access and digital IR as a supplement to more traditional IR.
- The EU has initiated a new wave of ESG legislation to increase the focus on the SDGs.
- Taxonomy is the overreaching classification framework to define what sustainable activities are.
- CSRD must be implemented in all major companies (listed and unlisted) from 1 January 2023, and likely all smaller companies as of 1 January 2026.
- SFDR must be implemented by all major investment companies (asset managers and private equity companies) from January 2022 and all investment companies from 1 January 2025.
- The amendments to MiFID II to encompass SFDR are applicable from August 2022 and latest by November 2022.

Part IV: Achieve a Fair Valuation of the Company Through Best-Practice IR

- Be present—an IR contact should always be available to call or call back shortly.
- Be proactive and seek to differentiate the IR function on good service, e.g. distribute relevant articles, distribute historical figures in new formats in good time, etc.
- Be careful not to become too good friends with investors or equity analysts, e.g. at roadshows. They are always in the market for extra information. Confidential information or hints must never be given to anyone.
- If the financial markets can still obtain a piece of the given information from another party, e.g. on the internet, from competitors or can even calculate the result, then give a helping hand and obtain the free goodwill.
- Listen to financial markets criticism and use it constructively.
- Do not blacklist equity analysts for negative views on the company.
- Perform an external IR advisor-based perception study every one or two years.
- Summary of good advice—best practice.
- Avoid surprising the financial markets, especially negatively.
- Proactively tell the financial markets about any problem areas.
- Be consistent in the company's communication and messages.
- Seek to minimise the time management spends on IR without this being perceived negatively in the financial markets. Instead, it gives extra time to run the business. But still, make sure management maintains good contact with investors and equity analysts.
- Do not damage the company's long-term development to satisfy the financial markets' desire for short-term results.
- Do not give in to temporary fads among capital market players if it is against the interests of the long-term business.
- IR is a strategic management tool. Therefore, IR's behaviour and information flow must be governed by the company's strategy—not in the short term by the financial markets' preferences.
- Be open and honest (within the framework of the law and to the extent permitted by competition).
- Be sure of the accuracy of all statements, and do not sweep anything under the carpet.
- Never tell investors or equity analysts anything which may not end up on the internet.
- Respond quickly to inquiries.
- Have the IRO give the board a brief update regarding the financial markets' view of the company before each board meeting.

- See what competitors are doing, including in the IR field, and learn from it.
- Providing structured monthly reporting to senior management and the board of directors with equity information not related to the financial performance of the company.

Part V: IR in Special Situations

- Always have up-to-date contingency plans and be ready in advance—before any issue arises.
- Have the work set in a fixed framework and ensure a periodic update of the emergency preparedness and a quarterly or semi-annual update of the company's defence bible.
- Let the work of special situations, including issue management and takeover risk, receive the necessary focus in the board of directors and the senior management.
- Ensure up-to-date white knight analysis and strategy plan in takeover scenarios
- Monitor market activity to ensure up-to-date defence bible and IMC to always be on top of emergency communication in unexpected situations.
- Appoint an internal operations manager to establish and maintain the emergency preparedness upon takeover. The IRO or CFO is recommended for this role.
- Ensure that the company can constructively challenge the company's external advisers, especially on the valuation side.
- Identify the issues that may attract investor activists' attention and deal with them.
- Ensure regular shareholder engagement and acknowledge that some of the contacts to the most important shareholders, which in some countries earlier solely rested with the CEO and CFO, today more appropriately takes place with the chairperson. In a potentially aggressive investor activist situation, a close early and historical relationship with the companies' major shareholders is key to winning their support rather than this support landing with the investor activist.
- If a possibility in respect to selective disclosure rules, consult the company's major shareholders before putting forward important shareholder proposals at the AGM or an EGM.
- Cater retail investors through digital IR as they often lack loyalty in investor activist situations.
- Ensure that the IR function monitors market activity, including strange movements and trading patterns, and keep an updated shareholder register. If the company identifies an investor building a significant shareholding, do not wait until the potential investor activist flags a 5% shareholding. Instead, reach out early for a dialogue.

- If approached by an investor activist, including a hedge fund activist, then embrace and listen to them. They are typically very knowledgeable, very well-prepared and very worthwhile listening to, and the company can typically learn a lot from them. It just requires confidence and an open mind.
- Considering the process of and IPO and the IRO's involvement before and after a company's listing.

Part VI: Embracing Non-financial/ESG Reporting

- Apply the UNGC's implementation framework on the UN's SDGs. On a board level, anchoring the ambition, thereafter, deepens the operational integration and lastly enhancing stakeholder engagement.
- Major companies, institutional investors and policymakers are embracing ESG, non-financial reporting and stakeholder capitalism at an unprecedented scale
- COVID-19 has elevated focus on ESG due to changes in consumer sentiment, workforce changes, increased legislation, health concerns and stakeholder awareness.
- The world is lagging behind the SDGs, and global initiates are taking place.
- Non-financial and ESG reporting will be a prerequisite for any company in navigating best-practice IR.
- There are clear advantages in non-financial reporting, and institutional investors are integrating these ESG-driven factors in their investment strategies.
- SFDR is a game-changer for institutional investors, which will put pressure on all institutional investors and, therefore, all companies.
- SFDR requires all investors to disclose their ESG strategy, performance and objectives on a fund level.
- The first detailed disclosure report on a product level must have been published by 1 January 2022 ("starting of the reference period"), then by 1 January 2023 (including comparison to the reference period), then by 1 January 2024 (including in comparison with the reference period and previous years) and so on. In 2027, the investment company must include a five-year development.
- Funds located in the jurisdictions of third countries outside the EU are also obligated to comply if they market towards EU investors.

Part VII: The Framework of Best-in-Class ESG Reporting

- Taxonomy is implemented in three stages: economic activity, eligibility and reporting KPIs. Remember to align the taxonomy framework with internal policies.
- At the 2022 Davos Summit, a panel participant on ESG reporting asserted that never had she witnessed so many abbreviations in one industry before. We have attempted to outline the official reporting frameworks here, which we believe companies can rely on. In the appendix, an overview of the Nasdaq's reporting guideline is found.
- The company must calculate their taxonomy eligibility across revenue, OPEX and CAPEX in their implementation of CSRD.

Part VIII: Preparing the Company's First ESG Report

- A key foundation for the first non-financial report: reporting boundaries, consolidation, period, performance and trends, accounting standards.
- The non-financial report should be implemented in four stages, using the E's of: establish, expand, embed and enhance.
- Apply a double materiality approach and conduct annual material analysis of the impact on multiple shareholders.
- Research and map out which ESG topics are considered material for the materiality assessment.
- Just like financial performance, you get what you measure. Align non-financial reporting with setting incentives throughout the organisation.
- Apply the SMART framework and integrate quantitative goals, qualitative milestones, relevant metrics, employee participation, applying weights, include periodic goals in non-financial performance.

Part IX: Aiming for Best-in-Class ESG Reporting

- Enhance non-financial reporting by being transparent and integrating financial with the non-financial reporting.
- Include a content index of the integrated reporting and the reporting standards.
- Mobilise the organisation and draft a non-financial reporting policy.
- Monitor trends and anticipate the direction of trace and participation of the various stakeholders, including the legislators.
- Utilise third-party ESG rating companies to conduct benchmarked and identify strengths and weaknesses in the company's approach to ESG.
- Embrace the new ESG-related role of equity analysts.
- Consider CSRD when embracing baseline non-financial reporting data.

- Include IR and perception studies to shape the non-financial reporting the expectations of the financial community and the company's stakeholders.
- Know your audience: Keep non-financial reporting simple and avoid complicated language and graphics.
- Beware of the overall best-practice IR communication and integrate non-financial reporting on the company's website, annual report and other IR tools.
- Establish relationships with the ESG rating agencies and expert ESG auditors.
- Beware of continuous benchmarking and perception studies.
- Make corporate access events dedicated to the equity story—but also from the perspective of ESG.
- Leverage technology for IR (e.g. digital IR) and information.

Part X: Future Trends of Financial and Non-financial Reporting

- Integrate consumer and retail investors in the strategic ESG reporting process.
- Consider digitalisation solutions to enhance reporting.
- Moving beyond GDP: include people, prosperity and the planet alongside profit.
- From a board of directors and senior management perspective, show caring, create trust and see the big pictures.
- Consider the role of the SDGs in corporate financing.
- Consider the megatrends in the long-term business strategy.

CHAPTER 44

Overview of Nasdaq's ESG Reporting Guidelines

Environmental Categories of Reporting

Due to the significant implications of climate change and the relatively quantitative approach to measuring CO_2 footprint, we believe environmental reporting is the most established form of non-financial reporting. We expect this to increase with the legislative trends of reducing emissions and protecting the planet. In the tables below, we outline the ten main categories of environmental reporting. The tables list the main categories, 1–3 sub-categories, how they are measured disclosed and their connection to the central regulatory frameworks explored in the previous chapter. It is also relevant for implementing taxonomy, which we explored in Chapter 34 (Figs. 44.1, 44.2, and 44.3).

Social Categories of Reporting

CSR and social justice have been a frequently discussed topic dating back decades, and the social categories are likely to be the most underrated of the ESG categories. In the tables below, we outline the ten main categories of social reporting. The tables list the main categories, 1–3 sub-categories, how they are measured, disclosed and connected to the central regulatory frameworks. The categories are a combination of quantitative and qualitative measures. As the categories are very sector-specific, a company needs to set its own targets and align them with their peers and the SDGs (Figs. 44.4, 44.5, and 44.6).

P. Lykkesfeldt and L. L. Kjaergaard, *Investor Relations and ESG Reporting in a Regulatory Perspective*, https://doi.org/10.1007/978-3-031-05800-4_44

Governance Categories of Reporting

Whereas the environmental and social categories of the ESGs have primarily been brought to the company's attention by external stakeholders, governance is a cornerstone of best business practices. Proper management, transparent disclosure and ethics are critical for a successfully listed company. As such, academics and investors exploring and deploying ESG strategies have long found the best correlations and causations between the category of governance (vs environmental and social) and the risk premium of equities. In the tables below, we outline the ten main types of governance reporting. In addition, the tables list the main categories, 1–3 sub-categories, how they are measured, how they are disclosed and their connection to the central regulatory frameworks (Figs. 44.7, 44.8, and 44.9).

1/3, Environment	GhG emissions	Emissions intensity	Energy usage
Sub-category 1	Total amount, in CO2 equivalents, for Scope 1 (if applicable)	Total GhG emissions per output scaling factor	Total amount of energy directly consumed
Sub-category 2	Total amount, in CO2 equivalents, for Scope 2 (if applicable)	Total non-GhG emissions per output scaling factor	Total amount of energy indirectly consumed
Sub-category 3	Total amount, in CO2 equivalents, for Scope 3 (if applicable)		
How is it measured?	By tracking the actual or estimated atmospheric emissions produced as a direct (or indirect) result of the company's consumption of energy	By dividing annual emissions (numerator) by various measures of economic output (denominator)	Typically measured in megawatt-hours (MWh) or gigajoules (GJ)
How is it disclosed?	As a number, trended over time (and compared against historical and industry averages, if possible)	As a number, trended over time (and compared against historical and industry averages, if possible)	As a number, trended over time (and compared against historical and industry averages, if possible)
Connection to frameworks	• GRI: 305-1, 305-2, 305-3 • UNGC: Principle 7 • SASB: General Issue / GHG Emissions (see also: SASB Industry Standards) • TCFD: Metrics & Targets (Disclosure B)	• GRI: 305-4 • SDG: 13 • UNGC: Principle 7, Principle 8 • SASB: General Issue / GHG Emissions, Energy Management (see also: SASB Industry Standards)	• GRI: 302-1, 302-2 • SDG: 12 • UNGC: Principle 7, Principle 8 • SASB: General Issue / Energy Management (see also: SASB Industry Standards)

Fig. 44.1 Overview of ESG categories—environment 1/3

2/3, Environment	Energy intensity	Energy mix	Water usage
Sub-category 1	Total direct energy usage per output scaling factor	Percentage: Energy usage by generation type	Total amount of water consumed
Sub-category 2			Total amount of water reclaimed
How is it measured?	By dividing annual consumption (numerator) by various measures of physical scale (denominator)	By quantifying the specific energy sources most directly used by the company	Water consumed, recycled and reclaimed annually, in cubic metres (m3)
How is it disclosed?	As a number, trended over time (and compared against historical and industry averages, if possible)	As a number, trended over time (and compared against historical and industry averages, if possible)	As a number, trended over time (and compared against historical and industry averages, if possible)
Connection to frameworks	• GRI: 302-3 • SDG: 12 • UNGC: Principle 7, Principle 8 • SASB: General Issue / Energy Management (see also: SASB Industry Standards)	• GRI: 302-1 • SDG: 7 • SASB: General Issue / Energy Management (see also: SASB Industry Standards)	• GRI: 303-5 • SDG: 6 • SASB: General Issue / Water & Wastewater Management (see also: SASB Industry Standards)

Fig. 44.2 Overview of ESG categories—environment 2/3

3/3, Environment	Environmental operations	Climate oversight / board	Climate oversight / Ex. Mgt.	Climate risk mitigation
Sub-category 1	Does your company follow a formal Environmental Policy? Yes/No	Does your board of directors oversee and/or manage climate-related risks? Yes/No	Does your senior management team oversee and/or manage climate-related risks? Yes/No	Total amount invested, annually, in climate-related infrastructure, resilience and product development.
Sub-category 2	Does your company follow specific waste, water, energy and/or recycling polices? Yes/No			
Sub-category 3	Does your company use a recognised energy management system? Yes/No			
How is it measured?	Companies that create, publish and periodically update a policy document that covers this subject may affirmatively respond	Companies that cover climate risk in board meetings (as part of the official agenda) or have a board committee dedicated to climate-related issues may affirmatively respond	Companies that cover climate risk in senior management meetings (as part of the official agenda) or have a management committee dedicated to climate-related issues may affirmatively respond	Companies measure the total dollar amount (USD) invested in climate-related issues, including R&D spend
How is it disclosed?	As text, with appropriate links to public content	As text, with appropriate links to public content	As text, with appropriate links to public content	As a number, trended over time (and compared against historical and industry averages, if possible)
Connection to frameworks	• GRI: 103-2 (see also: GRI 301-308 for relevant topic-specific standards) • SASB: General Issue / Waste & Hazardous Materials Management (see also: SASB Industry Standards)	• GRI: 102-19, 102-20, 102-29, 102-30, 102-31 • SASB: General Issue / Business Model Resilience, Systemic Risk Management (see also: SASB Industry Standards) • TCFD: Governance (Disclosure A)	• GRI: 102-19, 102-20, 102-29, 102-30, 102-31 • SASB: General Issue / Business Model Resilience, Systemic Risk Management (see also: SASB Industry Standards) • TCFD: Governance (Disclosure B)	• UNGC: Principle 9 • SASB: General Issue / Physical Impacts of Climate Change, Business Model Resilience (see also: SASB Industry Standards) • TCFD: Strategy (Disclosure A)

Fig. 44.3 Overview of ESG categories—environment 3/3 (*Source* Nasdaq [2019])

1/3, Social	CEO pay ratio	Gender pay ratio	Employee turnover	Climate risk mitigation
Sub-category 1	CEO total compensation to median FTE total compensation	Median male compensation to median female compensation	y-o-y %-chg for full-time employees	Total amount invested, annually, in climate-related infrastructure, resilience and product development
Sub-category 2	Does your company report this metric in regulatory filings? Yes/No		y-o-y %-chg for part-time employees	
Sub-category 3			y-o-y %-chg for contractors and/or consultants	
How is it measured?	As a ratio: The CEO Salary & Bonus (X) to Median FTE Salary, usually expressed as "X:1"	As a ratio: The median total compensation for men compared to the median total compensation for women	Percentage of total annual turnover, broken down by various employment types	Companies measure the total dollar amount (USD) invested in climate-related issues, including R&D spend
How is it disclosed?	As a number, trended over time (and compared against historical and industry averages, if possible) (S1.1); As text, with appropriate links to public content (S1.2)	As a number, trended over time (and compared against historical and industry averages, if possible)	As a number, trended over time (and compared against historical and industry averages, if possible)	As a number, trended over time (and compared against historical and industry averages, if possible)
Connection to frameworks	• GRI: 102-38 • UNGC: Principle 6 • Dodd-Frank regulatory guidance (US)	• GRI: 405-2 • UNGC: Principle 6 • SASB: General Issue / Employee Engagement, Diversity & Inclusion (see also: SASB Industry Standards)	• GRI: 401-1b • UNGC: Principle 6 • SASB: General Issue / Labour Practices (see also: SASB Industry Standards)	• UNGC: Principle 9 • SASB: General Issue / Physical Impacts of Climate Change, Business Model Resilience (see also: SASB Industry Standards) • TCFD: Strategy (Disclosure A)

Fig. 44.4 Overview of ESG categories—social 1/3

2/3, Social	Gender diversity	Temporary worker ratio	Non-discrimination
Sub-category 1	Total enterprise headcount held by men and women	Total enterprise headcount held by part-time employees	Does your company follow a sexual harassment and/or non-discrimination policy? Yes/No
Sub-category 2	Entry- and mid-level positions held by men and women	Total enterprise headcount held by contractors and/or consultants	
Sub-category 3	Senior- and executive-level positions held by men and women		
How is it measured?	Percentage of male-to-female metrics, broken down by various organisational levels	Percentage of Full-Time (or FTE-equivalent) positions held by non-traditional workers in the value chain	Companies that create, publish and periodically update a policy document that covers this subject may affirmatively respond
How is it disclosed?	As a number, trended over time (and compared against historical and industry averages, if possible)	As a number, trended over time (and compared against historical and industry averages, if possible)	As text, with appropriate links to public content
Connection to frameworks	• GRI: 102-8, 405-1 • UNGC: Principle 6 • SASB: General Issue / Employee Engagement, Diversity & Inclusion (see also: SASB Industry Standards)	• GRI: 102-8 • UNGC: Principle 6	• GRI: 103-2 (see also: GRI 406: Non-Discrimination 2016) • UNGC: Principle 6 • SASB: General Issue / Employee Engagement, Diversity & Inclusion (see also: SASB Industry Standards) • Guidelines for Multinational Enterprises, OECD'2011

Fig. 44.5 Overview of ESG categories—social 2/3

3/3, Social	Injury rate	Global health & safety	Child and forced labour	Human rights
Sub-category 1	Percentage: Frequency of injury events relative to total workforce time	Does your company follow an occupational health and/or global health & safety policy? Yes/No	Does your company follow a child and/or forced labour policy? Yes/No	Does your company follow a human rights policy? Yes/No
Sub-category 2			If yes, does your child and/or forced labour policy. See also: Cover suppliers and vendors? Yes/No	If yes, does your human rights policy. See also: Cover suppliers and vendors? Yes/No
How is it measured?	Total number of injuries and fatalities, relative to the total workforce	Companies that create, publish, and periodically update a policy document that covers this subject may affirmatively respond	Companies that create, publish, and periodically update a policy document that covers this subject may affirmatively respond	Companies that create, publish, and periodically update a policy document that covers this subject may affirmatively respond
How is it disclosed?	As a number, trended over time (and compared against historical and industry averages, if possible)	As text, with appropriate links to public content	As text, with appropriate links to public content	As text, with appropriate links to public content
Connection to frameworks	• GRI: 403-9 • SDG: 3 • SASB: General Issue / Employee Health & Safety (see also: SASB Industry Standards)	• GRI: 103-2 (see also: GRI 403: Occupational Heath & Safety 2018) • SDG: 3 • SASB: General Issue / Employee Health & Safety (see also: SASB Industry Standards)	• GRI: 103-2 (see also: GRI 408: Child Labour 2016, GRI 409: Forced or Compulsory Labour, and GRI 414: Supplier Social Assessment 2016) • SDG: 8 • UNGC: Principle 4,5 • SASB: General Issue / Labour Practices (see also: SASB Industry Standards)	• GRI: 103-2 (see also: GRI 412: Human Rights Assessment 2016 & GRI 414: Supplier Social Assessment 2016) • SDG: 4, 10, 16 • Universal Declaration of Human Rights, 1948 • UNGC: Principle 1, 2 • SASB: General Issue / Human Rights & Community Relations (see also: SASB Industry Standards)

Fig. 44.6 Overview of ESG categories—social 3/3 (*Source* Nasdaq [2019])

1/3, Governance	Board diversity	Board independence	Incentivised pay
Sub-category 1	Total board seats occupied by women (as compared to men)	Does company prohibit CEO from serving as board chair? Yes/No	Are senior managers formally incentivized to perform on sustainability? Yes/No
Sub-category 2	Committee chairs occupied by women (as compared to men)	Total board seats occupied by independents	
How is it measured?	The percentage of female directors and committee chairs, relative to male colleagues in the same groups	Companies with such a rule on the record may respond affirmatively; the number of "Independent Directors" (as defined in the board rules or corporate charter) as compared with other board members is also calculated	If senior managers are financially incentivised to perform on ESG metrics, the company may affirmatively respond
How is it disclosed?	As a number, trended over time (and compared against historical and industry averages, if possible)	As text, with appropriate links to public content (1); as a number, trended over time (and compared against historical and industry averages, if possible) (2)	As text, with appropriate links to public content
Connection to frameworks	• GRI: 405-1 • SDG: 10 • SASB: General Issue / Employee Engagement, Diversity & Inclusion (See also: SASB Industry Standards)	• GRI: 102-23, 102-22	• GRI: 102-35

Fig. 44.7 Overview of ESG categories—governance 1/3

2/3, Governance	Collective bargaining	Supplier cost of conduct	Ethics and anti-corruption
Sub-category 1	Total enterprise headcount covered by collective bargaining agreement(s)	Are your vendors or suppliers required to follow a Code of Conduct? Yes/ N	Does your company follow an Ethics and/or Anti-Corruption policy? Yes/No
Sub-category 2		If yes, what percentage of your suppliers have formally certified their compliance with the code?	If yes, what percentage of your workforce has formally certified its compliance with the policy?
How is it measured?	By measuring the number of employees governed by collective bargaining protocols against the total employee population	Companies that create, publish and periodically update a policy document that covers this subject may affirmatively respond	Companies that create, publish and periodically update a policy document that covers this subject may affirmatively respond
How is it disclosed?	As a number, trended over time (and compared against historical and industry averages, if possible)	As text, with appropriate links to public content (1); as a number, trended over time (and compared against historical and industry averages, if possible) (2)	As text, with appropriate links to public content (1); as a number, trended over time (and compared against historical and industry averages, if possible) (2)
Connection to frameworks	• GRI: 102-41 • SDG: 8 • UNGC: Principle 3 • SASB: General Issue / Labour Practices (see also: SASB Industry Standards)	• GRI: 102-16, 103-2 (see also: GRI 308: Supplier Environmental Assessment 2016 & GRI 414: Supplier Social Assessment 2016) • SDG: 12 • UNGC: Principle 2, 3, 4, 8 • SASB: General Issue / Supply Chain Management (see also: SASB Industry Standards)	• GRI: 102-16, 103-2 (see also: GRI 205: Anti-Corruption 2016) • SDG: 16 • UNGC: Principle 10

Fig. 44.8 Overview of ESG categories—governance 2/3

3/3, Governance	Data privacy	ESG reporting	Disclosure practices	External assurance
Sub-category 1	Does your company follow a Data Privacy policy? Yes/No	Does your company publish a sustainability report? Yes/No	Does your company provide sustainability data to sustainability reporting frameworks? Yes/No	Are the sustainability disclosures assured or validated by a third party? Yes/No
Sub-category 2	Has your company taken steps to comply with GDPR rules? Yes/No	Is sustainability data included in your regulatory filings? Yes/No	Does your company focus on specific UN Sustainable Development Goals (SDGs)? Yes/No	
Sub-category 3			Does your company set targets and report progress on the UN SDGs? Yes/No	
How is it measured?	Companies that create, publish, and periodically update a policy document that covers this subject may affirmatively respond	Does your company publish a sustainability report: Yes, No? If yes, the location of relevant public information should be declared. And does your company include ESG data in its regulator filings: Yes, No?	Does your company publish a GRI, CDP, SASB, IIRC or UNGC report? If yes, the location of relevant public information should be declared for each framework	Are your company's ESG disclosures assured or validated by a third party: Yes/No? If yes, please identify the audit/validation entity and the location of any relevant public information.
How is it disclosed?	As text, with appropriate links to public content	As text, with appropriate links to public content	As text, with appropriate links to public content	As text, with appropriate links to public content
Connection to frameworks	• GRI: 418 Customer Privacy 2016 • SASB: General Issue / Customer Privacy, Data Security (see also: SASB Industry Standards) • General Data protection Regulation (GDPR)	• UNGC: Principle 8	• UNGC: Principle 8	• GRI: 102-56 • UNGC: Principle 8

Fig. 44.9 Overview of ESG categories—governance 3/3 (*Source* Nasdaq [2019])

Correction to: The Formation of Stock Prices

Correction to:
Chapter 2 in: P. Lykkesfeldt and L. L. Kjaergaard, *Investor Relations and ESG Reporting in a Regulatory Perspective*, https://doi.org/10.1007/978-3-031-05800-4_2

The original version of this chapter was inadvertently published with incorrect Fig. 2.2 in Chapter 2, which has now been replaced with the correct figure. The chapter has been updated with the change.

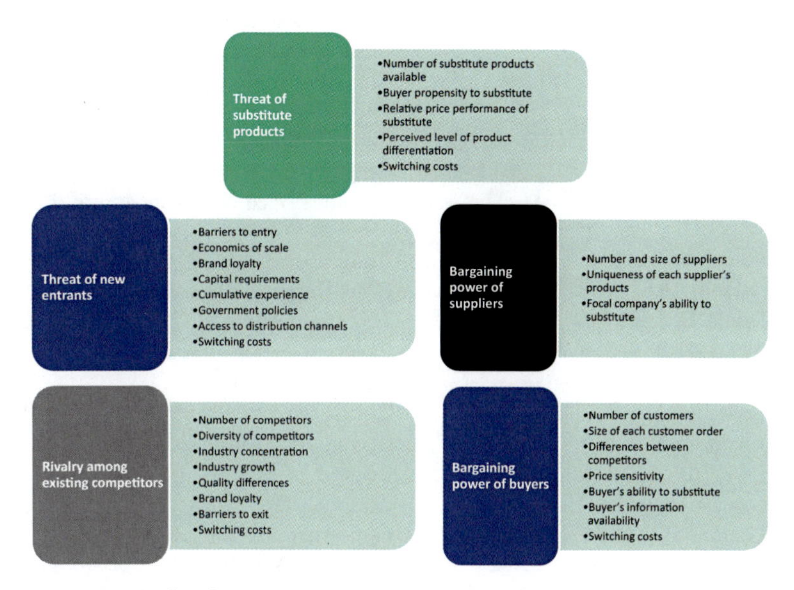

Fig. 2.2 Porter's five forces (*Source* Porter, M. [1979])

The updated version of this chapter can be found at
https://doi.org/10.1007/978-3-031-05800-4_2

P. Lykkesfeldt and L. L. Kjaergaard, *Investor Relations and ESG Reporting in a Regulatory Perspective*, https://doi.org/10.1007/978-3-031-05800-4_45

Index

P. Lykkesfeldt and L. L. Kjaergaard, *Investor Relations and ESG Reporting in a Regulatory Perspective*, https://doi.org/10.1007/978-3-031-05800-4

S

T

U

V

W